MINING FUNDAMENTALS

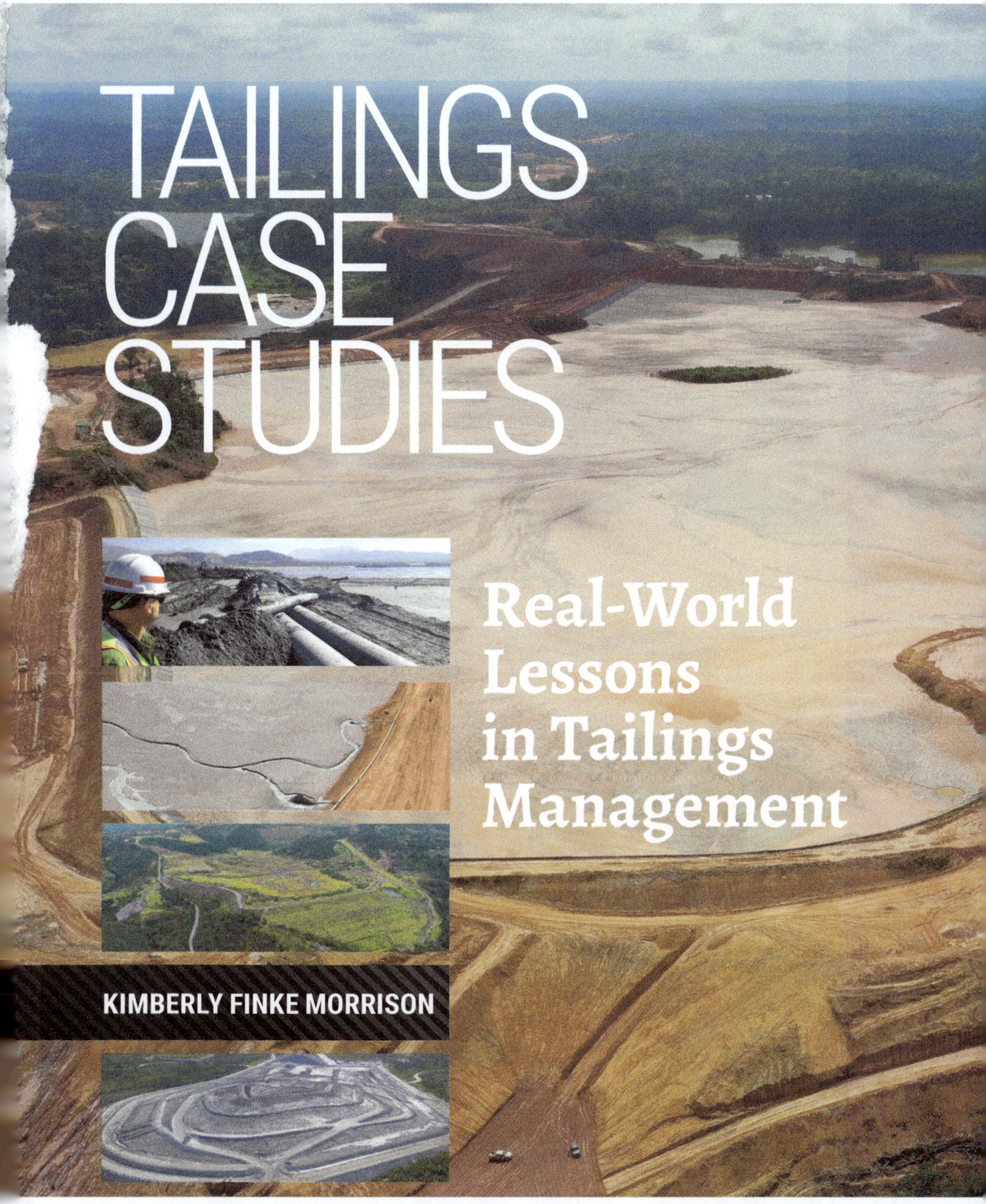

TAILINGS CASE STUDIES

Real-World Lessons in Tailings Management

KIMBERLY FINKE MORRISON

PUBLISHED BY THE SOCIETY FOR MINING, METALLURGY & EXPLORATION

Society for Mining, Metallurgy & Exploration (SME)
12999 E. Adam Aircraft Circle
Englewood, Colorado, USA 80112
(303) 948-4200 / (800) 763-3132
www.smenet.org

SME advances the worldwide mining and underground construction community through information exchange and professional development.

ISBN 978-0-87335-524-7
Ebook 978-0-87335-525-4

Library of Congress Cataloging-in-Publication Data has been applied for

Contents

Preface

Tailings management is a critical aspect of the mining industry. It involves the deposition and long-term storage of waste materials that are produced during the extraction of minerals from the earth. The safe and responsible management of tailings is essential to protect the environment and the communities that surround mining operations.

Over the course of the past several years, the mining industry has witnessed both remarkable successes and unfortunate failures in tailings management practices. The paradigm shift toward rapid and dramatic improvements in tailings management has been driven in part by the drive for industry to improve its environmental, social, and governance (ESG) practices, but the key driver for change has been tailings failure events. Failures and incidents are often the catalyst for change, and this is evident with tailings facility failures and the inherent increase in control. Development of the Global Industry Standard on Tailings Management (GISTM) by the Global Tailings Review in 2020 was an initiative driven by the failure of Dam 1 at Córrego do Feijão near Brumadinho, Brazil, a reaction based on lack of trust. It is through a deep understanding of these experiences—both positive and negative—that we can enhance our approach to tailings management and strive for continuous improvement.

In this book, we delve into the lessons learned from various case studies, highlighting the importance of sharing knowledge and experiences to drive positive change. While it is essential to acknowledge the failures and challenges faced in tailings management, this book primarily emphasizes successes from around the world, exploring the unique challenges faced, strategies employed, and outcomes achieved. By highlighting these accomplishments, we aim to inspire and encourage industry professionals to adopt innovative approaches and best practices, ultimately leading to improved global tailings performance. The case studies in this book are organized thematically and divided into five comprehensive categories as described next, which offer a structured approach to understanding the multifaceted aspects of tailings management.

DESIGN AND CONSTRUCTION

We begin with a focus on design and construction, exploring how innovative design practices have contributed to enhanced tailings performance and showcasing how emerging technologies are revolutionizing the field and improving overall performance. Several case studies are focused on improving tailings technology for existing tailings facilities, such as changes from conventional slurry deposition to thickened tailings deposition at the Musselwhite mine (Canada) or filtered tailings deposition at mines in Mexico and Colombia. The use of co-deposition of tailings with waste rock and heap leach ore are highlighted via case studies at the Éléonore (Canada) and Yanacocha (Peru) mines, respectively. The other case studies in this category discuss good practices with respect to geotechnical studies, analyses, and design.

OPERATIONS AND MONITORING

From there, we delve into the operational aspects of tailings management, examining successful strategies for efficient and sustainable management during mining operations, including monitoring practices, such as real-time monitoring at CBMM (Brazil), use of satellite imagery, as well as the use of a mobile geographic information system for improved documentation of inspections.

CLOSURE AND RECLAMATION

The book also examines the crucial phase of closure and reclamation, shedding light on effective approaches to safely and responsibly decommission and close tailings facilities. A new case study has been introduced providing a proposed approach to evaluation of safe closure, as introduced in the GISTM. Several tailings closure case studies are provided, including reclamation of the tailings facilities at San Xavier (United States), the Tanami mine (Australia), and the Marlin mine (Guatemala).

ENVIRONMENTAL AND REGULATORY

Environmental and regulatory considerations are becoming increasingly more of a focus when it comes to responsible tailings management. A critical aspect of building trust in tailings management is to ensure that the industry is held accountable for its actions. This means that mining companies must comply with relevant regulations and standards and be subject to independent monitoring and verification. This can help to ensure that tailings are managed safely and responsibly, and that any issues are identified and addressed promptly. Mining companies must also take steps to minimize the environmental impact of tailings management. Case studies include changes in the regulatory framework as a result of tailings facility failures in Brazil, deposition of high-sulfide tailings (Portugal), and activities to improve dust management from a tailings facility (Mexico).

LESSONS LEARNED FROM FAILURES

While the emphasis is on successes, we do not shy away from discussing past failures. By analyzing these failures, we can identify the root causes, learn from mistakes, and implement improved risk management measures to prevent similar incidents in the future. These lessons learned from failures serve as a reminder of the importance of continuous improvement and the need for robust and comprehensive tailings management practices. Case studies are provided from the tailings facility failures at Stava (Italy) in 1985, Omai (Guyana) in 1995, Cadia (Australia) in 2018, and Brumadinho (Brazil) in 2019.

CONCLUSION

In conclusion, this case study book on tailings management serves as a valuable resource for industry professionals, researchers, and policymakers alike. By sharing the successes and lessons learned, we aim to foster a culture of knowledge exchange, collaboration, and innovation in the field of tailings management. Together, we can build a safer, more sustainable, and more responsible future for the mining industry and the communities it serves.

—Kimberly Finke Morrison

About the Editor

As founding Chair of the Tailings and Mine Waste Committee of the Society for Mining, Metallurgy & Exploration (SME), formed in 2020, Kimberly (Kim) Finke Morrison, PE, RG, actively works to raise awareness about responsible tailings management through programs and training. Kim served as the managing editor for the first edition of SME's *Tailings Management Handbook: A Life-Cycle Approach*, published in early 2022, and has authored or coauthored nearly 40 other technical publications.

She is an active member of the International Council on Mining and Metals (ICMM)'s Tailings Working Group and a member of the Advisory Board for the Tailings and Industrial Waste Engineering Center (TAILENG). Formerly, she served in various leadership roles for the Geoprofessional Business Association (GBA), including 4 years on the Board of Directors (the second woman appointed to the board during its first 50 years), as chair of the Tailings Engineer of Record (EoR) Task Force, chair of the Education Committee, and co-chair of the Diversity and Inclusion Task Force.

For 5 years, commencing in 2019, Kim served as the global function lead for tailings, dams, and heap leach at Newmont's corporate headquarters in Denver, Colorado, where she was responsible for oversight of this critical infrastructure. In this role, she led development of standards, guidelines, and procedures; implemented and oversaw governance programs; led risk assessments; supported due diligence efforts; performed peer reviews; oversaw third-party reviews; developed training and provided mentoring to global team members; and was responsible for leading the company's implementation of the Global Industry Standard on Tailings Management (GISTM). Kim also served as the Chair of the Steering Committee for the GeoStable Tailings Consortium, which brings together eight global mining companies to study options to combine tailings and waste rock to create "geo-stable" landforms.

She received her bachelor's degree in Geological Engineering from the Missouri University of Science & Technology (formerly University of Missouri-Rolla) in 1996 and her master's degree in civil engineering-geosystems from the Georgia Institute of Technology in 1998. She is currently a registered Professional Engineer in five U.S. states and a Registered Geologist in one. She was a consultant for more than 20 years, including running her own engineering and environmental consultancy for over 5 years, before bringing that experience into the mining industry in late 2017.

In 2019, she received GBA's Award of Excellence in recognition of her meritorious leadership and achievement, culminating in the publication of GBA's *Proposed Best Practices for the Engineer of Record (EoR) for Tailings Dams*. In 2020, she was named to the *100 Global Inspirational Women in Mining* (WIM100) and received the SME President's Individual Citation Award for significant individual contribution and leadership in the mining industry. In 2022, she received the Environmental Stewardship Distinguished Service Award from SME's Environmental Division, recognized for the development, conservation, and protection of the earth's environment and natural resources through sustainable practices. In 2023, she was awarded the Jerry R. Bayless Alumni Merit Award from the Missouri University of Science & Technology and, in 2024, was selected to receive the designation of SME Fellow, Class of 2024.

Kim married husband J. Eric Morrison in 2002, and they reside in Lakewood, Colorado, with their two teenage children, son Evan and daughter Kendall.

Contributing Authors

Amanda Adams
Principal
Stantec
Denver, Colorado, United States

George Afriyie
Senior Dams and Tailings Engineer
Newmont Corporation
Denver, Colorado, United States

Ayobami Alao
Geotechnical Engineer
Freeport-McMoRan Bagdad
Arizona, United States

Henry Antonio Arapa Alvarez
Mechanical Engineer
Freeport-McMoRan
Arequipa, Perú

Manuel Aparicio
Principal—Tailings Management (RTFE)
Newmont Telfer Mine
Western Australia, Australia

Juan Carlos Tacca Apaza
Senior Project Engineer
Freeport-McMoRan
Arequipa, Perú

Nelson Roldan Aranibar
Tailings Operations Senior Supervisor
Freeport-McMoRan
Arequipa, Perú

Shaun Banks
Responsible Tailings Facility Engineer
Newmont Porcupine Mine
Timmins, Ontario, Canada

Brian Bellmore
Senior Civil Engineer
Barr Engineering Company
Duluth, Minnesota, United States

Johan Boshoff
Vice President Global Tailings Management
Gold Fields Limited
Perth, Western Australia, Australia

Nathan Brandner
Senior Geologist
Barr Engineering Company
Grand Rapids, Michigan, United States

Brett Byler
Director, Tailings and Dams
Newmont Corporation
Denver, Colorado, United States

Romeo Cabanayan
Senior Tailings Engineer
Newmont Red Chris Mine
British Columbia, Canada

Juan José Navarrete Cabrera
Mechanical Engineer
Freeport-McMoRan
Arequipa, Perú

José Campana
Geotechnical Consultant
Arcadis
Santiago, Chile

Terry Cutts
RDA Construction and Projects Superintendent
Newmont Australia
Boddington, Western Australia, Australia

V.V. Sosa Dávila
Geotechnical Engineer
Newmont Peñasquito
Mazapil, Zacatecas, México

Eduardo DeSousa
Principal Hydrogeologist
DHI Water and Environment
Perth, Western Australia, Australia

Scott Donald
Principal
WSP
Barrie, Ontario, Canada

Blair Douglas
Global Practice Lead, Water Mgmt./Hydrology
BHP
Perth, Western Australia, Australia

Leonardo Escott
Responsible Tailings Facility Engineer
Newmont Marlin Mine
San Marcos, Guatemala

Barry Esford
Manager, GeoEnvironmental Engineering
Cameco Corporation
Saskatoon, Saskatchewan, Canada

Michael Gabora
Director Geology and Hydrogeology
Lake Resources
Houston, Texas, United States

Jonathan H. Gillen
Principal Tailings Dam Engineer
Newmont Corporation
Denver, Colorado, United States

Philip Gomes
Senior Geotechnical Engineer
AtkinsRealis, Mining & Metallurgy
Montreal, Quebec, Canada

Ian Gordon
(Retired) Senior Tailings Advisor
Newcrest Mining Limited
Melbourne, Victoria, Australia

Ross Guenther
General Manager
Ceramext LLC
Reno, Nevada, United States

Ranjiv Gupta
Manager, Tailings, Crushed Leach, and Water
Freeport-McMoRan
Phoenix, Arizona, United States

Ross N. Hunsaker
Study Director
Newmont Corporation
Denver, Colorado, United States

Michael James
Engineer of Record
Agnico Eagle Mines Ltd.
Toronto, Ontario, Canada

Darrin Johnson
Associate Geotechnical Engineer
WSP
Mississauga, Ontario, Canada

Marcos Antônio Lemos Júnior
Senior Manager, Environmental & Dams Mgmt.
CBMM Araxá
Minas Gerais, Brazil

Art Kalmes
Vice President, Senior Civil Engineer
Barr Engineering Company
Minneapolis, Minnesota, United States

Homin Kim
Civil Engineer II
Freeport-McMoRan
Phoenix, Arizona, United States

Fannie Lessard
Mining Environment Engineer
AtkinsRealis, Mining & Metallurgy
Montreal, Quebec, Canada

Denise Levitan
Environmental Scientist and Geochemist
Barr Engineering Company
Minneapolis, Minnesota, United States

Ying Mao
Geotechnical Engineer
WSP
Mississauga, Ontario, Canada

Diego Bustinza Martinez
(Retired) Tailings Operations Senior Supervisor
Freeport-McMoRan
Arequipa, Perú

Seth McCann
Superintendent, Sustainability/External Relations
Newmont Corporation
Tanami Desert, Northern Territory, Australia

Kimberly Finke Morrison
Senior Director, Global Tailings Management
Newmont Corporation
Denver, Colorado, United States

James R. Obermeyer
Principal and Senior Vice President Emeritus
Stantec
Denver, Colorado, United States

Jean-François Painchaud
Responsible Tailings Facility Person
Newmont Éléonore Mine
James Bay, Quebec, Canada

Anamaria Rios Pando
Responsible Tailings Facility Engineer
Minera Yanacocha S.R.L.
Cajamarca, Perú

Krishna Parameswaran
President
tfgMM Strategic Consulting
Irving, Texas, United States

Pedro V. Porcayo
Principal Geotechnical Engineer
First Majestic Silver Corp.
Durango, Mexico

Humberto F. Preciado
Vice President, Mine Waste
WSP
Denver, Colorado, United States

E.J. García Romero
Responsible Tailings Facility Engineer
Newmont Peñasquito
Mazapil, Zacatecas, México

Daniel A. Servigna
Vice President, Mine Waste
WSP
Denver, Colorado, United States

Ben Sheets
Senior Geologist
Barr Engineering Company
Minneapolis, Minnesota, United States

João Antônio Teixeira Soares
Dams Supervisor
CBMM Araxá
Minas Gerais, Brazil

Mark Somppi
Director Processing
Newmont Corporation
Denver, Colorado, United States

Clint Strachan
Principal Geotechnical Engineer
Stantec
Fort Collins, Colorado, United States

Sharyn Thacker
Tailings Technical Specialist
Newmont Tanami Operations
Northern Territory, Australia

Robert Truby
Senior Engineer
Knight Piésold
Perth, Western Australia, Australia

Bryan Ulrich
Owner
Bryan Ulrich LLC—Tailings Solutions
Denver, Colorado, United States

Luis Valenzuela
Geotechnical Consultant
Honorary Member Institute of Mining Engineers
Santiago, Chile

Ryan L. Veenstra
Global Backfill Specialist
Newmont Corporation
Denver, Colorado, United States

Rens Verburg
Technical Fellow—Geochemistry
WSP USA
Redmond, Washington, United States

Michael Wallace
GIS Developer 2
Freeport-McMoRan
Phoenix, Arizona, United States

John Wates
Principal
John Wates Consulting
Rivonia, South Africa

Brant Whiting
Manager, Tailings, Crushed Leach, and Water
Freeport-McMoRan
Phoenix, Arizona, United States

David J. Williams
Professor of Geotechnical Engineering
The University of Queensland
Brisbane, Queensland, Australia

James L. Withiam
(Retired) Principal
D'Appolonia, Engineering Division of Ground Technology, Inc.
Johnstown, Pennsylvania, United States

Amir Hossein Zamani
Geotechnical Engineer
AtkinsRealis, Mining & Metallurgy
Montreal, Quebec, Canada

PART I

Design and Construction

Niobium Tailings Storage Facility Design: CBMM Dam 8, Minas Gerais, Brazil

Marcos Antônio Lemos Júnior
Companhia Brasileira de Metalurgia e Mineração (CBMM)

João Antônio Teixeira Soares
Companhia Brasileira de Metalurgia e Mineração (CBMM)

The Companhia Brasileira de Metalurgia e Mineração (CBMM) is a worldwide leader in the supply of niobium products, which can transform materials into intelligent solutions, building a more sustainable and efficient future.

CBMM's operation is in Araxá, State of Minas Gerais, in southeast Brazil. It comprises 15 industrial plants, including an ore processing phase. In this phase, the niobium concentrate is extracted from the ore and moves through the production flow, while the generated tailings are deposited as diluted pulp into the facility.

The tailings storage facility, which CBMM named "Dam 8," began operation in February 2019 and has an estimated life cycle of 8 years, considering the current production and market forecast. Dam 8 is a structure designed over compacted earth fill, with an internal drainage system, a reservoir lined with high-density polyethylene (HDPE) geomembrane, drains to collect water from springs and detect leakages, a surface spillway, as well as an industrial water-recycling system (Figure 1).

The embankment is raised using the downstream method on homogeneous compact earth fill. It was designed to be built with the mine waste material, thus avoiding the need to use new borrow pits, employing the existing materials in the operation of the ore-industry complex and

Courtesy of CBMM

FIGURE 1 Overview of the downstream region of Dam 8

maximizing the integration between the mining process and tailings deposition. The earth fill is raised in layers, with the layer thickness determined by experimental earth fills. To ensure that the project installation is in compliance, the company uses specialized technical follow-up to strictly monitor construction through field technological control and laboratory tests.

The dam fill also has an internal drainage system consisting of a vertical filter and horizontal blanket, built with sand and gravel. The purpose of the system is to collect naturally flowing water from springs in the area of the tailings facility, as well as eventual seepage that may occur from the reservoir, which is fully lined. The flow is directed to a concrete box with a V-notch weir, where the quantity and quality of the water is continuously monitored.

To avoid the tailings contacting the environment and maximize the reuse of the industrial water, the reservoir of the dam is fully lined with a 1.5-mm-thick HDPE geomembrane. This material undergoes strict quality control during its manufacture. To assure that all the properties of the geomembrane comply with technical specifications, after delivery to the construction site, samples are collected and sent to independent laboratories for verification.

During liner placement within the reservoir, the entire surface of the terrain is prepared by removing the vegetation layer and any granular material that may cause damage to the liner. Additionally, during the deployment, there is constant monitoring of the welds that connect the geomembrane rolls. The welds are made through heat fusion, ensuring the impermeability of the entire reservoir area (Figure 2).

A drainage system is installed under the liner system to collect and direct spring water in the reservoir area toward the downstream portion of the embankment, allowing its natural flow after the structure. Secondary drains are also built under the liner system to collect liner seepage and direct the flow downstream of the dam where it is collected and treated (Figure 3).

Surrounding the reservoir, a peripheral diversion channel was constructed, sized to convey flow from a 100-year storm event. The purpose of the channel is to reduce the flow directed toward the dam and divert it downstream of the structure. For rainfall events exceeding 100-year recurrence intervals, the excess flow is directed toward the reservoir of the dam, which has a spillway system sized to convey a storm event with a 10,000-year recurrence (Figure 4).

During the operation of the dam, the water from the reservoir is pumped by two mobile barges and conducted to the industrial plant. This process ensures a recirculation of 96% of the water needed for production. Rainfall dropping directly over the reservoir of the dam is also

Courtesy of CBMM

FIGURE 2 Deployment of the high-density polyethylene geomembrane

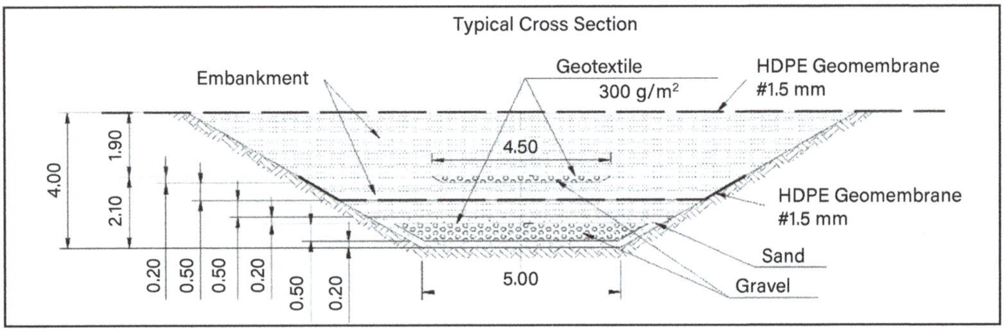

Courtesy of CBMM

FIGURE 3 Drainage system under the high-density polyethylene (HDPE) liner of the reservoir

Courtesy of CBMM

FIGURE 4 Drainage system in operation

captured and conveyed to a treatment station to comply with legal quality requirements before its release into natural water bodies.

The Dam 8 instrumentation plan was deployed to ensure that the behavior of the entire structure is monitored including the drainage system and any displacements of the embankment fill. All the necessary instruments to evaluate the safety of the dam are monitored full time through CBMM's Integrated Monitoring Center, which operates 7 days a week, 24 hours a day.

The process for tailings deposition performed by CBMM undergoes constant checks, performed by the company's technical team as well as by consultant experts, to ensure the safety of the structures and promote continuous improvement and the adoption of best practices. ■

Upstream Raise Tailings Storage Facility Design Considerations at Newmont's Tanami Operations, Northern Territory, Australia

Sharyn Thacker
Newmont Tanami

Newmont Tanami Operations (NTO) are located in the Tanami desert in the interior of Australia in the Northern Territory. It is a semi-arid environment with a monsoonal influence recognized for its diverse array of arid ecosystems, rare species, and intact habitats. Generally, evaporation rates exceed rainfall in all months of the year, and there are distinct wet and dry seasons. No permanent surface water sources exist, and the operation's water supply is sourced from groundwater aquifers.

The region is characterized by low topographic relief grading from low-lying sabkhas to highly weathered rocky outcrops and hills that impart limited impact on the landscape horizon. Cultural beliefs of the Warlpiri traditional owners are heavily connected to the land with songlines and dreamings that crisscross the landscape, often linked with unique landscape features.

NTO currently operates several paddock-style upstream raise tailings storage facilities (TSFs). Landforms constructed in association with mining activities have been limited to 15 m in height in agreement with traditional owner aspirations to limit visual impact on the landscape horizon. This low landform constraint reduces some of the risks typically associated with upstream raise TSFs (Figure 1).

The concept of tailings deposition at Tanami is based on pumping thickened tailings to the TSF at approximately 58% solids content. The tailings are discharged from the perimeter embankments of the TSFs to form a decant pond in the center of the cell. The tailings distribution pipeline includes spigot discharge points approximately every 40 m to evenly distribute tailings and enhance beach formation. Operating many spigots concurrently (four spigots on average) provides a steeper beach angle, which allows for better control of the centralized decant pond.

The tailings are allowed to settle, consolidate, and desiccate, with deposited tailings having less than 80% saturation, which makes the tailings stable for construction of subsequent upstream raises. The deposited tailings are dewatered above the segregation threshold providing a more uniform beach. The TSFs and/or cells are operated in rotation to enhance drying of the beach prior to subsequent construction (i.e., allowed to rest between construction cycles). The embankments are raised using construction materials borrowed from adjacent waste rock dumps.

Densities from the tailings thickener are maximized for the deposited tailings to optimize the storage capacity and, along with sufficient drying time, improve the tailings strength adjacent to the embankments to allow upstream raise construction. The design of the upstream

Courtesy of Newmont

FIGURE 1 Tanami's GTD08 (existing cells 1 and 2), Quorn, and Shoe in-pit tailings storage facilities

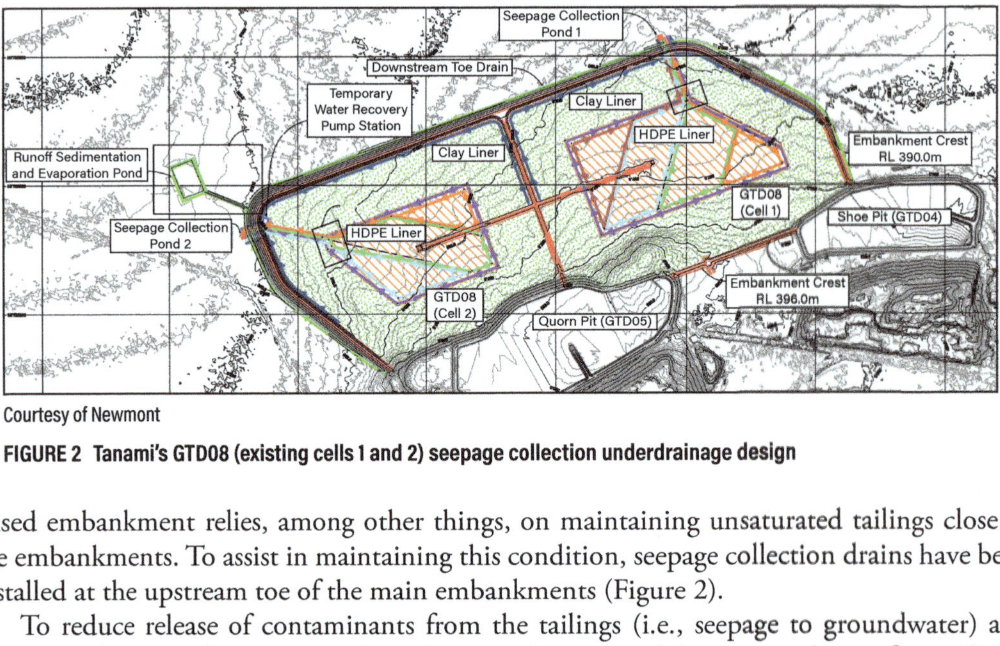

Courtesy of Newmont

FIGURE 2 Tanami's GTD08 (existing cells 1 and 2) seepage collection underdrainage design

raised embankment relies, among other things, on maintaining unsaturated tailings close to the embankments. To assist in maintaining this condition, seepage collection drains have been installed at the upstream toe of the main embankments (Figure 2).

To reduce release of contaminants from the tailings (i.e., seepage to groundwater) and minimize the groundwater mounding, the central section of the impoundment floor of each cell is equipped with a geomembrane high-density polyethylene (HDPE) liner, as well as an internal and cross-drainage system. Through collection and recovery from this seepage collection underdrainage system, the phreatic surface in the deposited tailings is generally maintained within the HDPE-lined area, thus minimizing seepage to groundwater.

The tailings deposition and water recovery systems have been designed so that after the initial start-up and tailings beach establishment, the decant pond can be manipulated to form at the center of the cell around the decant structure and away from the outer embankments.

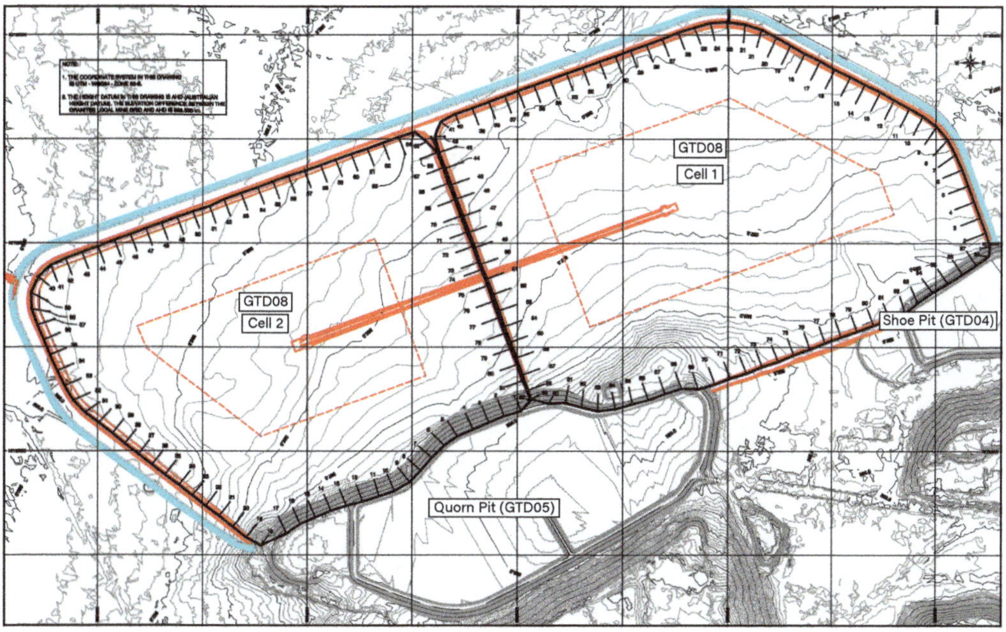

Courtesy of Newmont

FIGURE 3 Tanami's GTD08 (existing cells 1 and 2) perimeter spigot locations

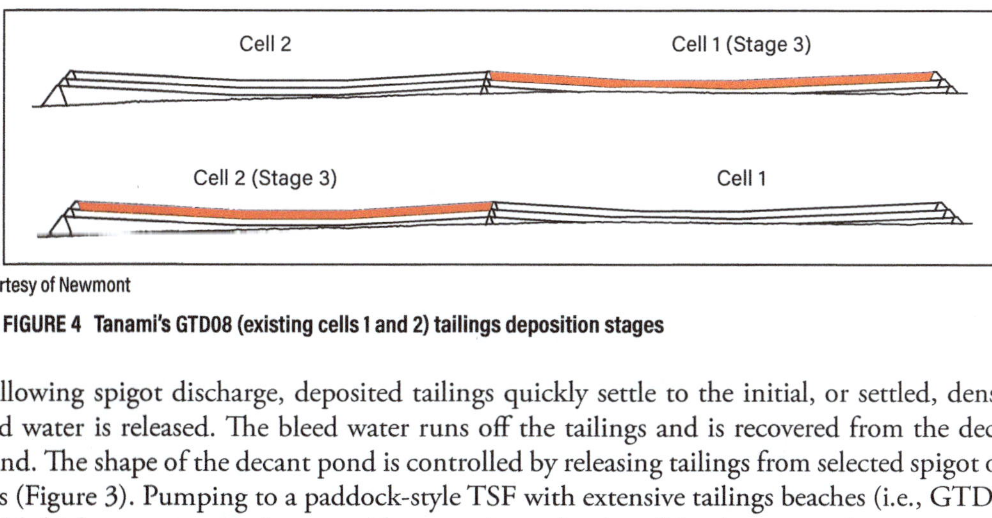

Courtesy of Newmont

FIGURE 4 Tanami's GTD08 (existing cells 1 and 2) tailings deposition stages

Following spigot discharge, deposited tailings quickly settle to the initial, or settled, density, and water is released. The bleed water runs off the tailings and is recovered from the decant pond. The shape of the decant pond is controlled by releasing tailings from selected spigot outlets (Figure 3). Pumping to a paddock-style TSF with extensive tailings beaches (i.e., GTD08) yields higher evaporation losses and hence reduces the size of the decant pond.

If tailings at the settled density are left to desiccate, they will reach the shrinkage limit, which is the water content at which a further loss of moisture will not result in a decrease in volume. In this state, the tailings achieve minimum volume and approach maximum strength. For tailings to reach the shrinkage limit, the deposition area must be large enough to ensure that placed tailings are not covered with fresh tailings before desiccation via evaporative drying is achieved (Figure 4). The area required to achieve this condition is calculated based on the evaporative potential at site, which for the Tanami TSFs is the minimum dry season evaporation rates. Also considered in the minimum required surface area are other influencing factors such as annual rate of rise, the constraints in the maximum embankment height, and the uncertainties in the future tailings characteristics and behavior. A conservative approach has therefore been adopted for NTO's TSFs, selecting a minimum cell surface area and allowing the tailings to approach maximum strength.

Upstream raise paddock-style TSFs have been assessed as a suitable design for the environmental conditions at the NTO site. No adjacent landforms for runoff water to enter the TSFs exist, and they are designed to store and contain the 10,000-year annual exceedance probability rainfall event. With high evaporation rates and long beaches, the tailings benefit from high drying potential. Additionally, the perimeter spigot operations and the embankment height limitations further support risk mitigation for these facilities. ■

Implementation of Passive Beach Drains, Boddington's F1/F3 Residue Disposal Area, Australia

Terry Cutts
Newmont Australia

Robert Truby
Knight Piésold Pty Ltd

George Afriyie
Newmont Corporation

The active tailings storage facility at Newmont's Boddington gold and copper mine, the F1/F3 residue disposal area, was commissioned in August 2009, and as of January 2021, the facility contains 400 Mt (million metric tons) of tailings. With a current mill processing rate of 40 Mt/y, the annualized rate of rise is 2.5 m.

The facility is confined by a main wall (saddle dam 1) across the principal catchment, which is currently raised in a centerline configuration, with additional saddle dams along the south, east, and north sides with natural topography along the western edge. The saddle dams are annually raised by the upstream method. The combined embankments cover a perimeter length of 12 km, with wall heights ranging from 25 to 55 m. Figure 1 shows the F1/F3 layout.

The upstream constructed embankments are not designed to be water-retaining structures. The design objective is to maintain a low phreatic surface within and near the embankments. The phreatic surface is controlled and kept low by an efficient underdrainage and seepage control system. The seepage and phreatic controls at the initial design and construction comprised:

- Toe drains located along the upstream toes of the upstream constructed dams
- Seepage cutoff trenches beneath the embankments
- Long tailings beaches to minimize the hydraulic gradient for seepage through and beneath the dams
- Underdrainage and leak collection recovery systems to extract water above and below the geosynthetic liner

As part of continuous improvement to assess embankment performance relative to the design intent, a formal review in 2018 highlighted the following:

- Drainage flows from the underdrainage network were less than design expectation.
- Piezometric readings from vibrating wire piezometers showed phreatic surface higher than the design expectation.
- Downstream of the tailings storage facility (TSF), groundwater levels were rising.

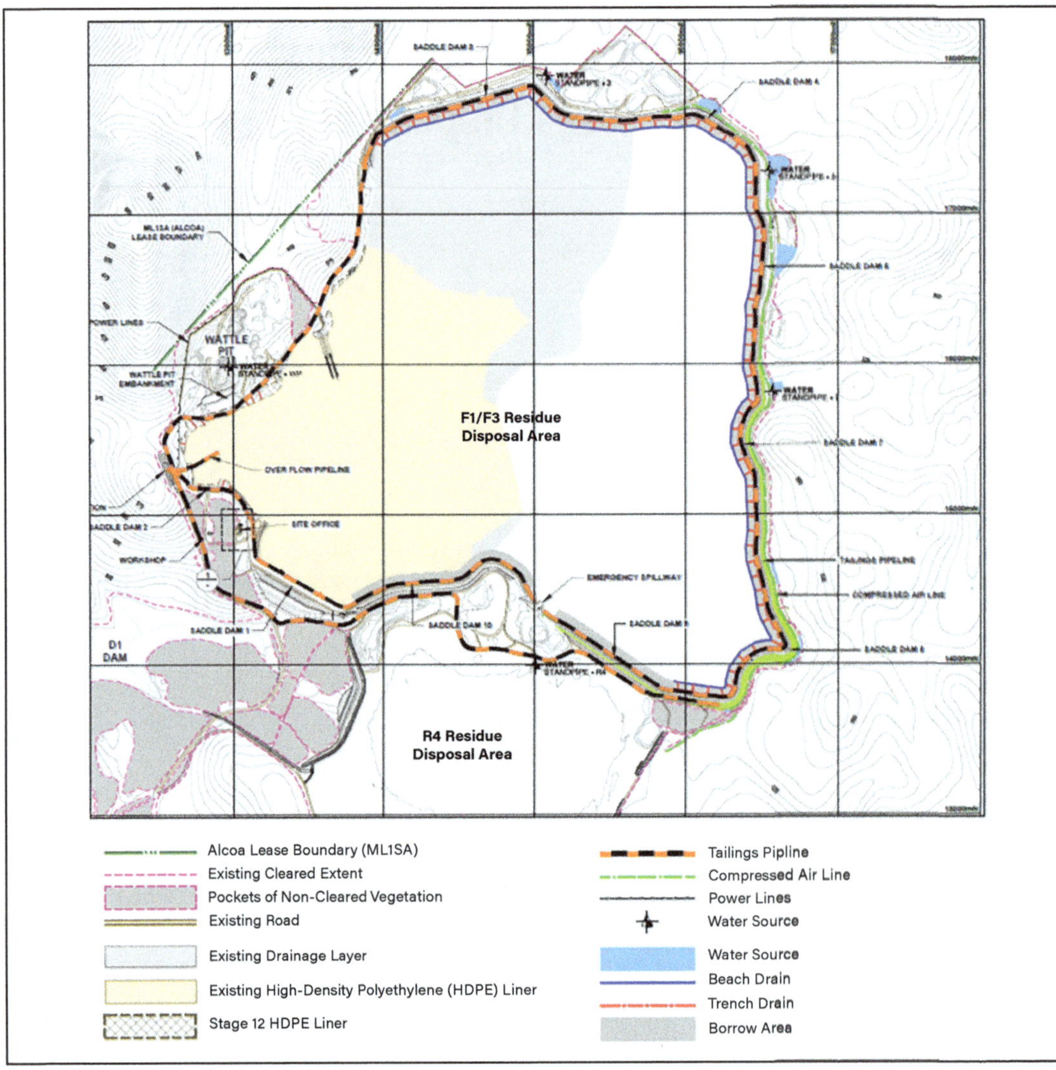

Courtesy of Newmont Corporation

FIGURE 1 F1/F3 tailings storage facility general infrastructure layout

A key outcome of this review was to upgrade the drainage design to enhance water extraction from the impoundment to mitigate further rise in the phreatic surface.

The upgraded drainage design included additional toe drains downstream of the embankments and beach drains on the upstream tailings beach along the entire length of the eastern and northern embankments as shown in Figure 1. The toe drains were excavated utilizing a Vermeer trencher to penetrate the lateritic hard cap to a depth of 900 mm with drain coil piping and filter-compatible gravels placed. The drains were all designed for minimum 1% gravity fall to adjacent surface perimeter sumps with the water then pumped via a constructed ring main pipe to a transfer point for use as return water to the process plant.

Before 2018, beach drains were installed along the upstream beach of the southern embankments (saddle dams 1, 10, and 9) with the drainage network reporting to six concrete towers placed in the tailings beach (Figure 2). The configuration of this beach drainage network causes several issues for operations including the following:

- Reliance on mechanical pumping for water extraction and transfer is impacted during annual raises when the towers are being raised and are offline for construction.
- The tower wells are fitted with pneumatic pumps, which have limited pump capacity to maintain low water heads in the drainage system and are maintenance intensive.

Courtesy of Newmont Corporation

FIGURE 2 Beach drain tower well and pumping infrastructure

In 2018, Newmont requested the design engineer to consider passive gravity drainage for future beach drains to provide a reliable cost-effective system with uninterrupted flow to maximize drainage efficiency and align with closure considerations.

This system utilized the previous method for the upstream beach drain installation, namely two lines of drainage with one close to the embankment and the other 100 m upstream (Figure 3).

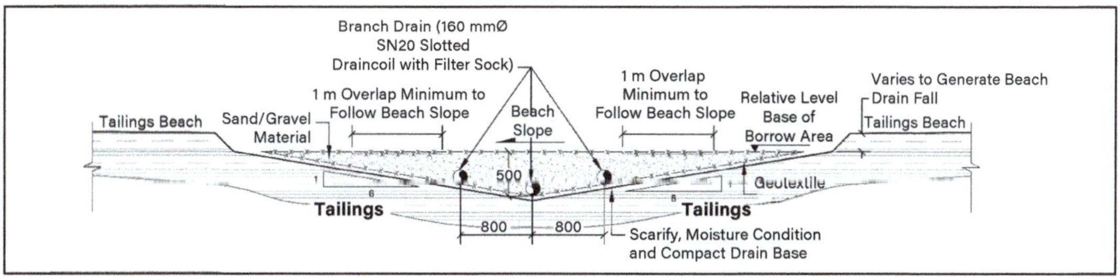

Courtesy of Newmont Corporation

FIGURE 3 Typical beach drain section within the upstream tailings beach

The passive drainage concept required excavation and placement of outfall pipes at approximately 350-m centers to connect the beach drainage to downstream discharge pipes (Figure 4). The pipe outfall detail comprises:

- Flexible outfall pipes
- Pipe-trench excavation through the existing embankment
- Pipe-trench foundation preparation
- A pipe-trench bedding layer
- Zoned pipe-trench backfill to reduce the risk of internal erosion, with
  - A low-permeability clay (zone A1) upstream zone
  - A 1-m-thick bentonite clay zone embedded within the low-permeability clay zone

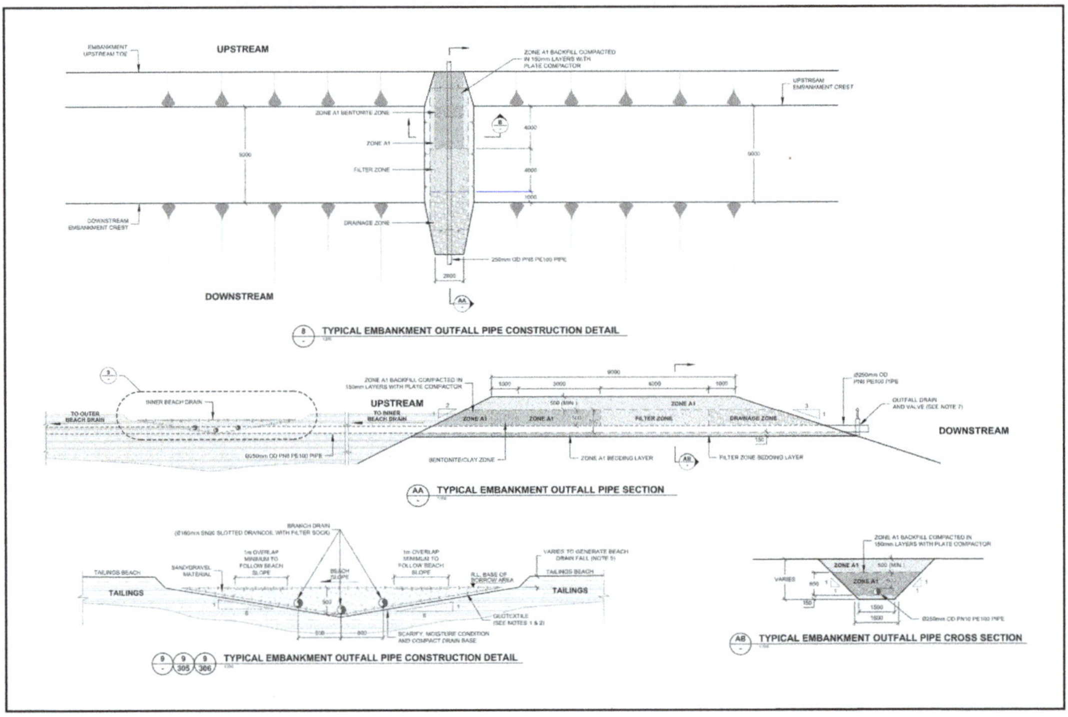

Courtesy of Newmont Corporation

FIGURE 4 Outfall pipe construction layout

- A filter zone, compatible with the tailings, downstream of the low-permeability clay zone
- A gravel drainage layer downstream of the filter zone

The construction of the passive drainage was undertaken during stage 12 (2018/2019) and stage 13 (2019/2020) TSF construction and was fully commissioned by mid-2020. Construction photos are shown in Figures 5 and 6.

Courtesy of Newmont Corporation

FIGURE 5 Backfilling upstream beach drain with filter-compatible gravels prior to encapsulation with geotextile

Courtesy of Newmont Corporation

FIGURE 6 Bedding layer and outfall pipe (verifying designed gradient)

Courtesy of Newmont Corporation

FIGURE 7 Beach drain outfall pipe (saddle dam 7 downstream) connected to discharge line

Courtesy of Newmont Corporation

FIGURE 8 Beach drain collector pipe and outlets at saddle dam 8

The drainage system was installed over a 6,900-m embankment length with 18 outlets, each connected to a continuous pipeline on the downstream toe for the gravity discharge. Refer to Figures 7 through 9.

The beach drain has proven to be highly effective in capping phreatic level rise within the upstream embankments. Recorded flow rates range between 5 and 40 m^3/h as shown in Figure 10.

Courtesy of Newmont Corporation

FIGURE 9 Water discharge at the downstream transfer point (saddle dam 9 sump)

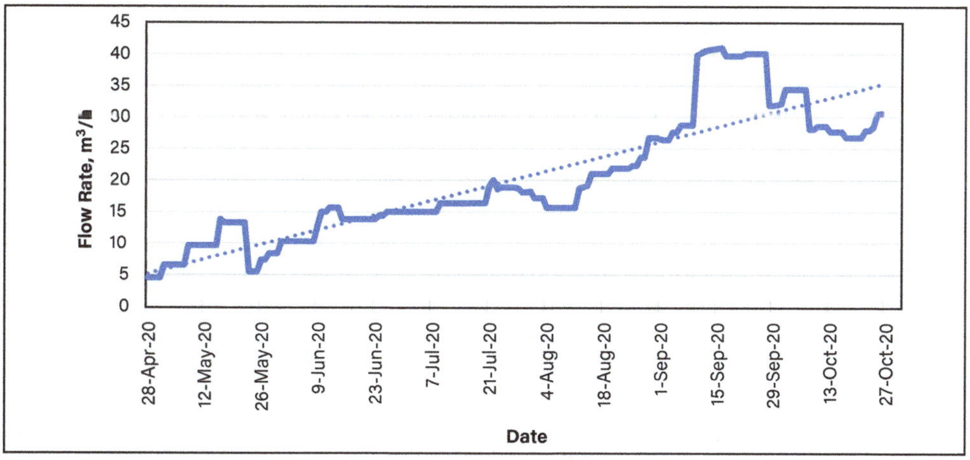

Courtesy of Newmont Corporation

FIGURE 10 Beach drain flow rate

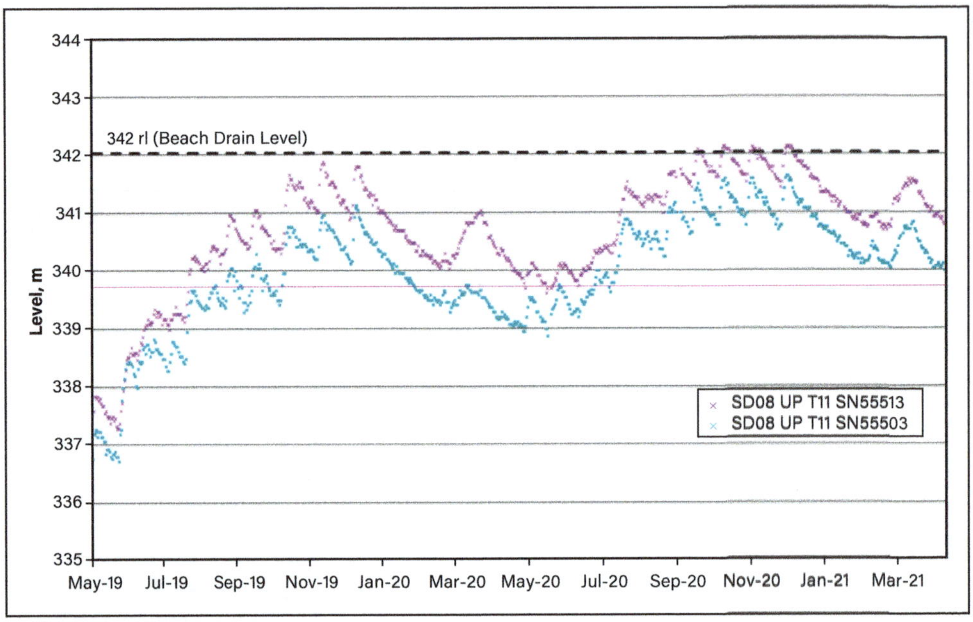

Courtesy of Newmont Corporation

FIGURE 11 Upstream piezometer levels (saddle dam 8)

Piezometric readings taken upstream, which previously increased annually during tailings deposition, are capped at the beach drain elevation as illustrated in Figure 11.

The beach drainage installed upstream connected to the gravity outfall pipes has proven to be a better solution to the towers previously installed. The passive system reduces reliance on mechanical pumping, reduces operational cost, and meets closure considerations. ■

Cerro Verde Expansion Project, Cerro Verde Mine, Arequipa, Peru

James R. Obermeyer
Stantec

Amanda Adams
Stantec

The collaborative design and construction process for a new tailings storage facility (TSF), as part of the Cerro Verde production unit expansion (CVPUE) project at the Cerro Verde mine, which took place from 2008 to 2015, is a success and one that can be replicated throughout the mining industry. The close partnership between the designer/engineer of record (EoR) Stantec (formerly MWH) and the mine owner and operator Sociedad Minera Cerro Verde S.A.A. (SMCV) and its ultimate parent company Freeport-McMoRan Inc. (FCX) facilitated a robust design and efficient project execution. FCX, SMCV, and Stantec continue to partner on the operational construction to rigorously achieve design expectations.

The Cerro Verde mine is located approximately 30 km south of the city of Arequipa, Peru. The mine has been in operation since the early 1970s and is currently producing both copper and molybdenum concentrates and copper cathodes. When the CVPUE project was developed, operations consisted of two open pits, a solvent extraction and electrowinning plant and associated heap leach facilities, a 120,000-t/d (metric tons per day) concentrator plant, and the associated Quebrada Enlozada TSF (Enlozada TSF). The Enlozada TSF, which was commissioned in 2006, was designed for a total capacity of almost 1 Gt (gigaton). The Quebrada Linga TSF (Linga TSF) was a major component of the Cerro Verde CVPUE project. The Linga TSF is designed to store 2.025 Gt of tailings generated at a rate of 240,000 t/d.

In June 2010, the Linga TSF location was selected as the best alternative for the CVPUE project. The selection of the site was followed by an embankment alternatives study. The centerline sand embankment was identified as the preferred embankment type based on the excellent performance of the Enlozada TSF, water efficiency, and experience of the mine operators with this construction method.

Involvement of the external technical review board began during design of the Enlozada TSF. The review board's involvement continued through the design, construction, and start-up of the Linga TSF and is still ongoing. During the design of the Linga TSF, presentations to the review board took place at key milestones and the board made a minimum of one site visit per year.

The Linga TSF consists of a zoned rock-fill starter dam that is raised by the centerline construction method, the tailings impoundment located upstream, and a seepage collection sump located downstream of the TSF embankment, with a two-stage cyclone station, reclaim water system, and tailings transport and delivery systems.

The 165-m starter dam was designed to provide storage for the first 2 years of operation. The TSF is to be raised an additional 140 m by constructing an embankment of compacted cycloned tailings sand using the centerline method. The ultimate height of the Linga TSF will be 305 m at the dam centerline, and the tailings impoundment will cover an area of approximately 17,300 ha (hectares).

Tailings from the new concentrator are thickened to a solids content of 55%, and the thickener underflow (whole tailings) is then transported through tailings delivery pipelines to the Linga TSF. The material balance analysis demonstrated that 50% of the whole tailings material would need to be cycloned to provide sufficient quantities of cyclone sand to support ongoing construction of the TSF embankment. The remaining 50% of the whole tailings is deposited in strategic locations in the TSF impoundment to manage the location and size of the reclaim water pond. The produced cyclone underflow sand is deposited in 30-cm lifts on the 3.5H:1V slope embankment and mechanically compacted.

Water that drains from the hydraulically deposited tailings underflow sand is collected in a system of embankment underdrains, conveyed to the seepage collection sump, and pumped back to the cyclone station for reuse. The seepage collection system drainage network was designed to enable the rapid drainage, transport, and recovery of water contained in the underflow sand and to maintain a low phreatic surface in the starter dam and sand embankment during tailings deposition. The seepage collection system drainage network will be expanded as the embankment grows during ongoing TSF operations.

The construction schedule was defined by the date that construction was permitted to start (May 2013) and date when the new concentrator commissioning was scheduled to occur (September 2015), but the starter dam construction had to be completed in August 2015 to allow time for jacking header construction and pipe installation on the starter dam crest by September. This provided a total of 27 months to complete activities for the starter dam, which included foundation excavation, grouting, material processing and production, and embankment fill placement. The crest of the starter dam reached the design height of 165 m on August 8, 2015. The total volume of starter dam fill placed was 13,730,000 m^3. Figure 1 shows the Linga TSF near the end of starter dam construction.

Courtesy of Stantec

FIGURE 1 Linga tailings storage facility near the end of starter dam construction

Construction engineering and quality assurance/quality control (QA/QC) during construction were critical to confirm that the construction complied with design intent and construction specifications and to quickly respond to contractor questions and

Courtesy of Stantec

FIGURE 2 Linga tailings storage facility after 5 years of operation

unanticipated conditions without delaying construction. Construction occurred 24 hours a day, 7 days per week, and resulted in on-time completion of the starter dam and other components of the TSF by the September 2015 concentrator commissioning date.

Concentrator start-up for the CVPUE project began as planned on September 1, 2015. During start-up, sand production and placement steadily increased from about 7,000 t/d at the start in 2015 to 29,000 t/d on average in mid-2016. Use of key performance indicators to monitor actual performance with planned performance was an important factor in the success of the first 2 years of operation and resulted in the compacted sand reaching the starter dam crest slightly ahead of time, on September 1, 2017, when the impounded tailings level in the TSF impoundment was 8 m below the crest.

At the end of December 2020, the elevation of the sand embankment reached a height of 209 m at the embankment centerline (2,614 mamsl [meters above mean sea level]). The height is measured from the base of the starter dam vertically to the lowest point of the embankment crest (the height measured from the downstream toe is greater than along the centerline of the embankment). The Linga TSF is considered to be highly successful and is performing as anticipated in the design. The Linga TSF is shown in Figure 2 after 5 years of operation.

Inspections and oversight by the EoR are ongoing. QC is performed by SMCV and QA activities and operations support are provided by Stantec around the clock. The FCX tailings stewardship team and the review board also annually inspect the facility. ■

Rock-Fill Dam Construction at a Tailings Storage Facility, Bagdad Mine, Arizona, United States

Ayobami Alao
Freeport-McMoRan

Ranjiv Gupta
Freeport-McMoRan

OVERVIEW

The Bagdad mine is located in west central Arizona, United States. Modern milling operations began at the Bagdad mine in 1977. Currently, the Upper Mammoth tailings storage facility (TSF) is the active TSF at the Bagdad mine. Upper Mammoth TSF is a centerline tailings dam with an underflow sand embankment. The ultimate height will be 126 m with a storage capacity of 517 Mt (million metric tons). The tailings deposition at Upper Mammoth TSF began in December 2013.

PROJECT BACKGROUND

To continue tailings deposition at Upper Mammoth TSF, a containment boundary to bridge the gaps in the natural topography on the east and west was required. Based on a multiple account analysis, a centerline rock-fill Upper Mammoth secondary dam (UMSD) was selected as the best option for the containment boundary. UMSD will be constructed to an ultimate height of 90 m in six 15-m lifts. The UMSD project started in 2018 and is anticipated to be completed in 2029. Figure 1 shows the topography of Upper Mammoth along with the east and west "pockets" of the rock-fill dam.

DESIGN

UMSD has design slopes of 1.4H:1V on the downstream side of the section and 1.5H:1V on the upstream side. The transition zone, where the underflow sand from the tailings dam overlaps with the rock-fill dam at the downstream side, has variable slopes ranging from 3:1 to 2.5:1 to 2:1.

CONSTRUCTION MATERIAL

UMSD is constructed with quartz monzonite, a waste rock generated from Bagdad mine operations. The dam includes a two-stage filter zone made up of select rock-fill material and fine-grained material that allows the passage of water but prevents tailings infiltration to the rock-fill layer.

Courtesy of Freeport-McMoRan

FIGURE 1 Upper Mammoth (UM) dam showing east and west pockets of the rock-fill dam

CONSTRUCTION METHODS

The main rock-fill section is constructed in 1-m lifts using bulldozers and compacted with a smooth drum vibratory roller. The rock fill is moisture conditioned to help obtain a homogeneous mixture for adequate compaction. Select rock fill is placed in 1-m-wide and 0.3-m-high lifts and compacted to ensure uniform density. Filter material is placed in 1.5-m-wide and 0.3-m-high lifts. Figure 2 shows ongoing rock-fill and filter construction at the dam.

QUALITY CONTROL

The owner's tailings operations team is constructing the dam. The quality control (QC) activities are performed by the owner's engineer and include daily construction observation to confirm the construction conforms to the design. The daily QC activities involve the assessment of material lift depth during placement, rock quality and gradation assessment, visual inspection, and documentation of the construction.

QUALITY ASSURANCE

In addition to QC activities, independent quality assurance (QA) testing is conducted by the design engineering team. The QA team visits the site at least once a month and reviews ongoing construction and documentation. The QA team also approves and documents foundation preparation in construction observation reports.

(A) Rock-fill construction

(B) Filter construction

Courtesy of Freeport-McMoRan

FIGURE 2 Upper Mammoth secondary dam rock-fill and filter construction

SUMMARY

The owner's team and the design team regularly communicate about any design or construction method changes. The requests for design or construction method changes are communicated and documented by the owner's team through request for information and implemented after approval by the design team. The successful construction of the UMSD involves close and careful coordination between the owner's team and the design engineering team. ■

Conversion from Conventional Slurry to Thickened Tailings Deposition, Musselwhite Mine, Ontario, Canada

Ying Mao
WSP

Mark Somppi
Newmont Musselwhite Mine

Romeo Cabanayan
Newmont Red Chris Mine

Darrin Johnson
WSP

This case study provides a summary of key considerations that were evaluated during a tailings storage capacity expansion option evaluation for Newmont's Musselwhite gold mine in Northwestern Ontario, Canada. Musselwhite commenced production in 1997 with tailings placed in a surface tailings management area (TMA) with conventional tailings discharged at a slurry density of 50% solids (by weight) and a production rate of approximately 4,000 t/d (metric tons per day). The site is in a region characterized by low seismicity and cold temperate climate. The studies outlined herein led to a corporate decision to change the method of deposition from conventional slurry to thickened tailings.

The Musselwhite TMA was initially designed with a series of engineered earth embankments, providing containment across topographic lows around the facility. The perimeter embankments incorporate a low-permeability central core that is keyed into the foundation. The starter embankments were constructed in 1996 and subsequently raised in 2000 and 2005 to store 17.3 Mt (million metric tons) of tailings with provision for a shallow water cover upon closure. However, by the early 2000s the projected mine resource had increased to 34 Mt, which significantly exceeded the original TMA design capacity and prompted the evaluation of expansion options.

TMA expansion trade-off studies began in 2002. Options that were considered included TMA footprint expansion, construction of a new facility, and alternate tailings technologies (i.e., increased levels of tailings dewatering). Construction of a new TMA was determined to be uneconomical and less environmentally favorable (i.e., from a water management and permitting perspective). Expansion of the existing TMA by raising the perimeter embankments to allow for continued conventional tailings deposition was determined to be costly and have increased safety risk because of a higher tailings pond level. The preferred option involved vertical raising of the existing TMA footprint by stacking thickened tailings that would not require constructing additional embankment raises or raising of the tailings pond elevation.

Construction of a thickened tailings stack above the deposited beach could provide the required tailings storage capacity. The tailings pond level can be maintained during operation, while it can be lowered at closure to facilitate a dry cover to replace the original water cover associated with conventional tailings discharge, thus reducing embankment failure risks. In summary, converting the TMA to a thickened tailings operation was viewed to have the following advantages over continued conventional slurry tailings deposition:

- Decreased tailings pond size, reduced seepage from the TMA, reduced risk of perimeter embankment failure, and improved overall water management (i.e., less water in TMA)
- Smaller overall footprint and active deposition area to allow progressive closure
- Steeper tailings beach slopes to allow for an improved capacity versus raise requirements (i.e., reduced perimeter embankment dam construction)
- Production of non-segregating tailings, which were expected to be relatively impervious, helping inhibit tailings oxidation and reduce the likelihood of acid generation and metals leaching in the long term, while reducing the potential for dust generation in the short term
- More conducive to reclamation and aesthetically more compatible to the surrounding landscape during closure
- Simplified dike raise construction (i.e., homogeneous granular internal dikes vs. zoned embankments)

Two concepts for thickened tailings deposition were then considered: (1) central discharge using one or more towers; and (2) perimeter discharge from internal raised dikes. Deposition using the central discharge concept would result in a cone-shaped facility, which would require more sophisticated deposition planning to maintain a radial tailings discharge pattern and avoid trapping water (i.e., maintain internal perimeter drainage). In addition, internal dike raising for a central discharge would have to be carried out on the wet toe of the cone (i.e., with more difficult construction and foundation issues).

Discharge of thickened tailings from the western perimeter of the TMA was the method adopted because it would create a beach with runoff draining toward the center of the TMA. This eliminates the risk of trapping water around the perimeter and potential earthworks and water management challenges. Internal tailings deposition dike raises are constructed at the top of the tailings beach where it is drier, resulting in a better foundation.

For the implemented perimeter discharge arrangement, thickened tailings are discharged from the western end of the TMA. As the tailings surface rises, the west discharge dike is progressively raised, and the north and south dikes are extended using the upstream construction method (Figure 1). Each deposition dike raise is 2 m in height and constructed with engineered granular fill. On the south side of the TMA, the internal discharge dike is offset

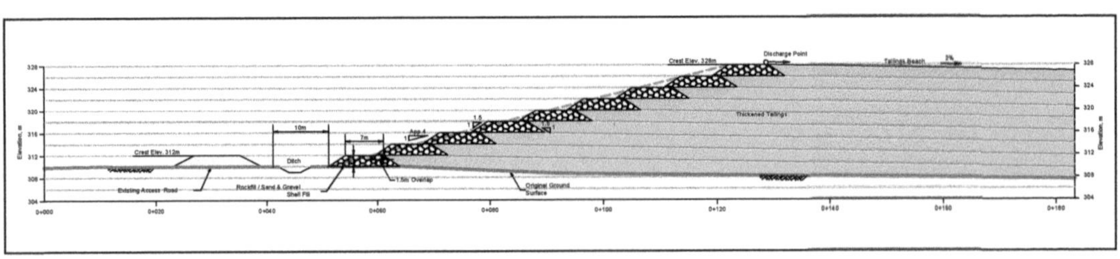

Source: Kam and Li 2008

FIGURE 1 Stacking of thickened tailings

Photograph by Shiu N. Kam 2013

FIGURE 2 Musselwhite tailings management area after conversion to thickened tailings (looking east)

50 m upstream from the perimeter embankments. This buffer zone was designed to provide containment of the thickened tailings stack in the event of a dike failure.

Comprehensive material characterization was completed to evaluate the properties of the thickened tailings. Bench test and field pilot test results indicated that the tailings could achieve an underflow density of 70% solids (by weight).

Preparation for thickened tailings deposition in the TMA began in 2007 with the construction of a separation dike partitioning the TMA into an east cell and a west cell. Thickened tailings deposition is carried out in the west cell, while the east cell continues to act as a water management pond (Figure 2).

Expansion of the Musselwhite TMA was carried out by converting the facility from a conventional slurry discharge operation with a water cover at closure to a thickened tailings deposition with a dry closure cover and a smaller pond. The tailings thickening plant was constructed in 2009 and ultimately commissioned in May 2010. For more information on this case study, readers are referred to Kam et al. (2011, 2014). Operations for both the TMA and the thickener have been satisfactory without major issues in the past decade. ■

REFERENCES

Kam, S.N., and Li, A.L. 2008. *Design of Thickened Tailings Discharge Dyke and Separation Dyke, Musselwhite Tailings Management Area, Musselwhite Mine.* Report No. 07-1118-0037. Ottawa, Ontario, Canada: Golder.

Kam, S., Girard, J., Hmidi, N., et al. 2011. Thickened tailings disposal at Musselwhite mine. In *Paste 2011: Proceedings of the 14th International Seminar on Paste and Thickened Tailings.* Edited by R.J. Jewell and A.B. Fourie. Perth: Australian Centre for Geomechanics.

Kam, S., Hmidi, N., and Mao, Y. 2014. Field performance of thickened tailings disposal at Musselwhite mine. In *Paste 2014: Proceedings of the 17th International Seminar on Paste and Thickened Tailings.* Edited by D. van Zyl, R.J. Jewell, and A.B. Fourie. Perth: Australian Centre for Geomechanics.

Conversion of a Tailings Storage Facility from a Cycloned Embankment to Earth Fill

Bryan Ulrich
Bryan Ulrich LLC

At a gold mine in southwest Colorado, a sidehill cycloned tailings storage facility (TSF) was converted to have an earth-fill embankment. This conversion was made to eliminate the often cumbersome cyclone operation and to improve the overall environmental performance of the facility, especially with respect to the airborne dust that could affect the nearby community. Because of land constraints, the downstream toe of the TSF could no longer advance outward. The engineered solution for this was to develop a design where the embankment partially stepped out onto the tailings beach.

A site investigation, including the cone penetration test, standard penetration test, self-boring pressure meter tests, along with drilling and sampling was completed, followed by a campaign of geotechnical tests conducted on the recovered materials. After a series of geotechnical analyses (including a liquefaction assessment, and slope stability and seepage analyses), it was established that the facility would benefit from the construction of an elevated beach drain, which, based on those analyses, was designed to keep the phreatic surface at a safe distance from the embankment. In the previous design concept, the highly permeable sandy embankment had maintained the phreatic surface in an acceptable position. However, in the revised design concept, with the embankment partially stepping out onto the tailings beach, this additional drainage feature was required.

The blanket drain was constructed along the crest of the facility, spanning some 46 m out onto the beach. Pipework was installed to direct and gravity drain collected water to an adjacent process water pond. Following completion of the drain, earth fill from a nearby borrow area was placed over the downstream slope of the TSF, covering the cycloned sand slope,

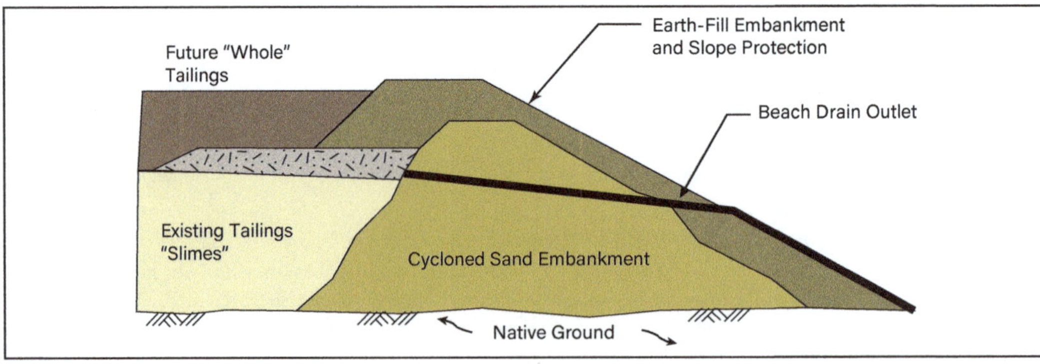

FIGURE 1 Tailings storage facility design

which previously could cause particulate pollution in the nearby community. The same type of earth fill was used to construct a raise for the embankment. The construction was completed during the summer season, and the facility operated as intended until its ultimate closure was completed. The conversion of the cycloned TSF to an earth-fill facility benefited the TSF operation by removing the cycloning operation and benefited the community by reducing the impact from dust.

In addition, the conversion of this facility to an earth-fill structure, including the placement of earth fill on the downstream slope of the TSF, led to the possibility of concurrent reclamation of the facility's side slopes, further protecting the nearby dwellings from windblown dust, and improving the quality of the scenic viewshed (Figure 1). ■

Successful Conversions of Former Slurry Tailings Embankments into Filtered Tailings Storage Facilities, Durango, Mexico

Humberto F. Preciado
WSP

Daniel A. Servigna
WSP

Pedro V. Porcayo
First Majestic Silver Corp.

This case study presents successful conversions of two conventional slurry tailings storage facilities (TSFs) to filtered tailings storage facilities in Durango, Mexico, for two mining operations with production rates ranging from 1,200 to 4,500 t/d (metric tons per day). In both cases, the decision to make the conversion was driven by:

- Ability to increase the storage capacity via stacking of filtered tailings
- Reduction of the tailings' impacted footprint
- Mitigation of long-term geotechnical stability risks
- Alleviation of environmental risks by reducing tailings water as a source of pollutants, and a net reduction in process make-up water demand

The case studies can be divided in two classes: Class 1 facilities, where slurry tailings are loaded almost immediately with filtered tailings because of storage area limitations; and Class 2 facilities where the conversion is immediate, but the available area allows gradual or eventual loading of the previously stored slurry tailings.

The condition where slurry tailings must be loaded almost immediately represents the greatest engineering challenge because of the geotechnical risks of instability and liquefaction (under both static and dynamic loads). Regardless of whether these facilities are converted immediately or eventually, they all require monitoring of pore water pressure conditions within the slurry tailings being loaded as the depth of placement of the filtered tailings above the slurry tailings increases. Additionally, where loading rate exceeds the pore pressure dissipation rate, tailings dewatering may be required to increase the rate of consolidation and strength and mitigate the risks of foundation failure.

GUANACEVÍ TAILINGS STORAGE FACILITY CONVERSION

The Guanaceví facility conversion falls within Class 1, where land availability limitations required the previously stored slurry tailings to be loaded almost immediately with filtered tailings. Because of local regulations in Mexico, and the use of cyanide leaching in the process at the Guanaceví mine (Endeavour Silver Corp.), the conventional TSF was geomembrane lined and lacked any drainage system. The conventional TSF prior to the conversion is shown in Figure 1.

Courtesy of Wood

FIGURE 1 Original Guanaceví conventional tailings storage facility

The original condition of the conventional TSF allowed a maximum placement of just a few meters of filtered tailings before the geotechnical factor of safety dropped below the minimum required in the design criteria or potentially triggered instability of the facility because of static liquefaction. Further analyses also indicated that piezometric pressure conditions of the slurry tailings had to be consistently lowered as the filtered tailings stack was being raised.

Given the importance of drainage for this conversion, various passive and active dewatering options were considered to reduce the in situ tailings pore water pressure and allow for the loading of the slurry tailings deposits. The passive options included vertical wick drains and horizontal drains. Active options included the removal or dredging of the slurry tailings, and the construction of an extraction pumping well network. Because of the need to load the existing slurry tailings as soon as feasible, and because of the opportunity to recover metal-rich solution from the tailings pore water, the decision favored the installation of a network of extraction wells early in the filtered tailings placement phase as soon as a thin cap of drain material or filtered tailings would facilitate equipment access. Figure 2 shows the converted TSF.

Courtesy of Wood

FIGURE 2 Converted Guanaceví filtered tailings storage facility

CUPÍAS TAILINGS STORAGE CONVERSION

The Cupías facility belongs to Class 2, where the available area allows gradual or eventual loading of the previously stored slurry tailings. This slurry TSF originally consisted of a closed valley embankment built using an upstream raise construction method. Currently, the Cupías TSF is owned and operated by First Majestic Silver Corp. Unlike Guanaceví, the flotation mineral processing allowed for an unlined TSF that could drain through the foundation, as well as through decant towers placed within the tailings impoundment (Figure 3).

Courtesy of Wood

FIGURE 3 Original Cupías conventional tailings storage facility

To improve the stability of the tailings embankment and achieve the maximum required storage capacity, stabilization berms (or buttresses) were designed at the downstream toe of the TSF. During design, it was determined that at least three stabilization berms would be required. Given the volumes of material necessary for the construction of these berms/buttresses, consideration was given to changing the tailings processing technology from conventional slurry to filtered tailings.

Once a decision was made to convert the conventional TSF to a filtered TSF, a portion of the planned buttress at the toe of the TSF was constructed with filtered tailings, which resulted in approximately 10 years of storage life before the slurry tailings within the existing impoundment had to be loaded. This period allowed for some drainage and consolidation of the previously deposited slurry tailings to facilitate loading by construction equipment and eventually loading by the filtered tailings stack reducing risk of liquefaction. Figure 4 shows the current configuration of the Cupías TSF with the buttresses completed and progressively reclaimed and filtered tailings being placed on top of the former slurry tailings impoundment.

Courtesy of First Majestic Silver Corp.

FIGURE 4 Condition of the Cupías tailings storage facility through the years after conversion to filtered tailings

CONCLUSIONS

Although these case studies apply to production rates below 10,000 t/d, advancements in filtering technology and co-deposition of various types of waste and dewatered tailings will make conversion of existing conventional TSFs a more common practice. In addition to the technical challenges, the cases presented in this study required the education, understanding, and commitment from the mine personnel at various levels from operations and construction to be able to implement a filtered tailings stack construction plan that was not prescriptive, but dependent on the evolution of site conditions and the level of dewatering achieved in the underlying slurry tailings impoundments. This level of interaction is key in the successful conversion of a conventional TSF to a filtered TSF. In both cases, storage capacity has more than doubled because of the successful conversion to filtered tailings management. ■

Converting from Slurry to Filtered Tailings Deposition in a Tropical Climate, Antioquia, Colombia

Daniel A. Servigna
WSP

Boris Garcia
Independent Consultant

This case study presents the successful conversion of a conventional slurry tailings storage facility (TSF) to a filtered tailings storage facility (FTSF) in the Department of Antioquia, central northwest part of Colombia.

Since the mid-19th century (1852), gold has been exploited from the Segovia-Remedios mining district, approximately 180 km southwest of Medellin, the capital city of Antioquia, Colombia. The mine operation under discussion is located within this mining district and produces approximately 1,800 t/d (metric tons per day) of tailings with a target production of 2,000–2,200 t/d. Before 2010, this mine was exploited by different owners and operators, with an average tailings production of 600 t/d, which was primarily discharged to a nearby creek (Figure 1).

Operational improvements were made at the mine to cease the discharge of tailings into the creek between 2010 and 2017. As of 2017, tailings were managed via conventional TSF. About 90% of the tailings containing negligible cyanide content were disposed of directly in the TSF, whereas the remaining 10% of the tailings leached with cyanide were further treated before disposal at the TSF.

The tailings generated by the process plant were pumped to designated dewatering pits for dehydration before being truck-hauled to the TSF for final deposition (Figures 2 and 3).

Photos by B. Garcia

FIGURE 1 Slurry tailings discharge directly to a local creek

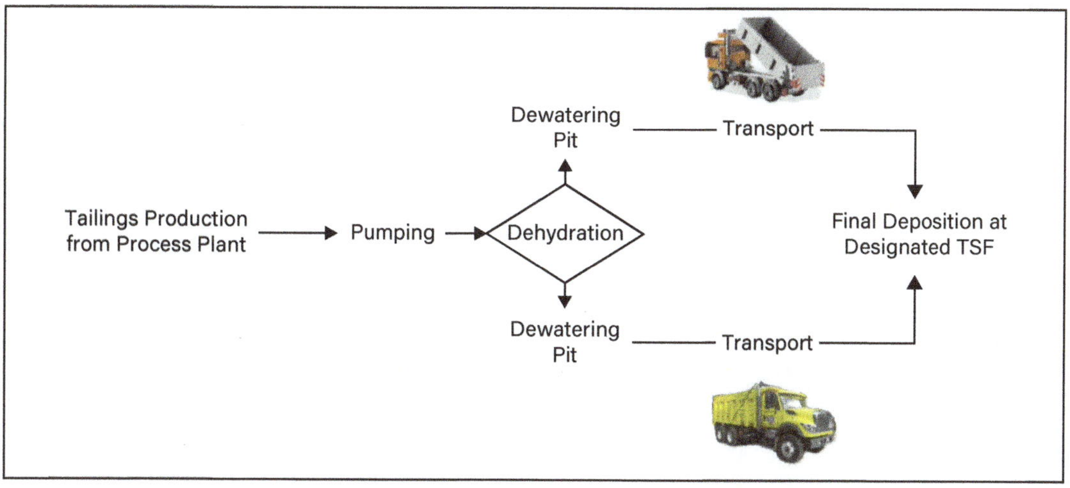

FIGURE 2 General schematic of tailings management circuit with final deposition at the tailings storage facility (TSF)

Photo by B. Garcia

FIGURE 3 Dehydrated tailings truck-hauled from one of the dewatering pits to the tailings storage facility

TOWARD SUSTAINABLE AND RESPONSIBLE TAILINGS DEPOSITION AND MANAGEMENT

Between 2012 and 2016, several studies were completed to evaluate the best option to dispose of tailings in compliance with local regulations and international best practices. Additional considerations included identifying opportunities for future expansions and low-cost closure. Among the tailings disposal technology options evaluated, conversion from slurry to filtered tailings was considered a favorable long-term solution when considering tailings mineralogy, chemistry, and other physical properties. However, significant precipitation was identified as a potential obstacle because of the climatic conditions in the region. Additional challenges included the following short- and medium-term aspects:

- The facility was in noncompliance with permissible limits for contaminants in the effluents from the dewatering pits, which included total suspended solids (TSS) and total cyanide (CN) content.
- The remaining storage capacity for tailings was limited at the two designated disposal sites, which provided less than 3 months of operational storage.

- The anticipated construction schedule for the initial stage of converting a slurry TSF to FTSF would exceed the available commissioning time frame.
- For long-term operation requirements, tailings storage capacity would need to be addressed.

EVALUATED OPTIONS FOR SHORT-, MEDIUM- AND LONG-TERM TAILINGS DISPOSAL

A comparative evaluation of options for immediate/short-, medium- and long-term solutions were evaluated. In all cases, both flotation and cyanidation streams were treated separately.

Immediate and Short Term

Two options were evaluated for short-term (ST) disposal:

1. ST1. Flotation tailings are dewatered in two of the existing dewatering pits, while cyanidation tailings are dewatered in a new separate pit. Under this approach, the flotation tailings stream—without cyanide content—would exhibit a higher settling rate at the dewatering pits; hence, it is expected that the resulting effluent's TSS will decrease. Alternatively, the cyanide tailings stream—in much lower flow rate—will be dewatered separately, while its effluent is treated by geotextile tube (e.g., Geotube) plus activated carbon package. The final effluent would be recirculated to the cyanidation stage, which potentially replaces the use of freshwater makeup in the process. Both dewatered tailings (~20% moisture) are also handled separately within the final disposal site in the final TSF, a closer location from the process plant than other proposed TSF locations.
2. ST2. The plan is similar to ST1 but considers additional techniques to enhance dewatering at the pits. These techniques are, for example, in-line flocculation and using dewatering well pumps in the pits.

Medium Term

Three medium-term (MT) options were evaluated:

1. MT1. Flotation tailings are pumped to a cyclone station in the TSF. The resulting underflow sand is used for dike construction while overflow is contained upstream. Cyanide tailings are dewatered in one of the existing dewatering pits, where their effluent is treated by Geotube plus activated carbon package. Cyanide dewatered tailings are finally disposed in dedicated cells at the TSF. The other existing (two in total) dewatering pits will be used as emergency dump areas.
2. MT2. Flotation tailings are pumped to a filtration plant at TSF area, where the resulting cake is disposed within the TSF. Cyanide tailings are dewatered in one of the dewatering pits, where their effluent is treated by Geotube plus activated carbon package. Cyanide dewatered tailings are finally disposed in dedicated cells at the El Chocho TSF. The other existing (two in total) dewatering pits will be used as emergency dump areas.
3. MT3. Flotation tailings are pumped to Geotubes for deposition at the TSF. Cyanide tailings are dewatered in one of the dewatering pits, where their effluent is treated by Geotube plus activated carbon package. Cyanide dewatered tailings are finally disposed in dedicated cells at the TSF. The other existing (two in total) dewatering pits will be used as emergency dump areas.

Long Term

Two long-term (LT) options were evaluated:

1. LT1. The option considers two dedicated tailings filtration plants. One of them will be fed by flotation tailings, while the other will be fed by the residue from the cyanide destruction circuit at the process plant. Both cakes are finally disposed in the TSF or another dedicated site for filtered tailings deposition.
2. LT2. Tailings are to be used as backfill within the underground mine workings. This option is similar to LT1, but additionally, it considers a paste backfill plant as an alternative to support mining activities. The backfill paste plant will use filtered tailings (product of the filter plant option LT1) and portland cement as binder to form a paste. The paste fill is delivered to underground workings either by gravity flow or by pumping. The cured cemented fill will provide structural support, allowing stope filling and ore pillar extraction consequentially.

RISK-BENEFIT ANALYSIS

Potential risks for all options were evaluated concerning their possible impact on the environment, public access roads, constructability, and technology. Additionally, qualitative potential benefits of each option were identified.

All of the proposed options implicated improvements to the current tailings handling system and assured compliance with the mine's objectives for the TSS and CN content of effluents. Additionally, the proposed options presented the following benefits:

- By implementing activated carbon/Geotube packages before discharge to the environment, there is the potential for gold recovery from exhausted effluents.
- Dewatering times in the pits will be reduced from 12 days to less than 7 days (estimated).
- By increasing the dewatering rates, there is a possible increased throughput at the process plant.
- Reduction of dewatering periods from 12 days to less than 7 days (estimated) will provide the capability to take in deposition at the designated TSF according to the plan outline for ST1 and ST2.
- Storage capacity would be increased at the existing TSF.
- Because dewatered tailings will no longer be hauled through the municipalities of Segovia and Remedios, Antioquia, noise and dust will be reduced as well as damage to the infrastructure and public roads.
- Emergency dump facilities will become available at zero cost (for the medium- and long-term options).

IMPLEMENTED SOLUTION

In 2018, the mine operator began the implementation of Options MT3 and LT1. While preparing the existing TSF area for early deposition of tailings, different conceptual FTSF start-up concepts, such as cycloning, filtration, and Geotubes, were implemented to maximize the overall storage capacity of the facility. During the FTSF design, special attention was given to the management of runoff. Two separate sets of underdrain and diversion channels were designed to divert non-contact water from the TSF area, and to store, treat, and safely discharge the contact runoff. Production of FTSF operations commenced in 2019, a year after the Geotubes were implemented at the facility (Figure 4).

In 2020, tailings deposition using the Geotube technology ceased, and the mine converted its tailings management deposition technology to filtered tailings disposal methods. A filtered tailings deposition allowed for the closure of the TSF areas that were previously used to store slurry tailings, as shown in Figure 5.

Photos by B. Garcia

FIGURE 4 Tailings slurry storage using Geotubes. Filtered tailings deposition began a year after the use of the Geotubes.

Reclaimed portion of the original tailings storage facility area

Tailings storage facility after 3 years of filtered tailings deposition

Photos by B. Garcia

FIGURE 5 Aerial views of portions of the original tailings storage facility

CONCLUSION

The conversion from disposing slurry tailings to disposing tailings in a dehydrated/filtered state results in an immediate improvement of the stability of the deposit, significantly reducing the risk of collapse of any TSF structure. The use of new technologies, such as the implementation of Geotubes and tailings filtration, resulted in a lower environmental impact on soil, air, water, flora, and fauna resources. The longer the life of the FTSF, the less land is needed to store tailings; and therefore, the removal of vegetation and its associated fauna is not required.

By using tailings dehydration/filtration technologies combined with equipment such as filter presses and geotextile tubes, the deposit increased its capacity, allowing at least 3 more years of storage plus the chance to be expanded for additional storage. Moreover, FTSFs can be progressively reclaimed, lowering closing costs in the long term. ■

Integrated Waste Rock Co-Deposition at the Éléonore Mine Tailings Storage Facility to Improve Geotechnical Stability and Facilitate Site Drainage, James Bay, Quebec, Canada

Jean-François Painchaud
Newmont Éléonore Mine

Philip Gomes
AtkinsRealis, Mining & Metallurgy

Amir Hossein Zamani
AtkinsRealis, Mining & Metallurgy

Fannie Lessard
AtkinsRealis, Mining & Metallurgy

PROJECT BACKGROUND

The Éléonore mine is located in the James Bay area, in Northern Quebec, about 800 km north of Montreal. The Éléonore mine consists of an underground gold mine, mill, water treatment plant, waste rock stockpile, tailings storage facility (TSF), access road, landing strip, and camp for up to 600 people. The mine has been in operation since October 2014, and its cessation is scheduled for 2026.

The TSF was designed with a 25 Mt (million metric ton) storage capacity. Each year, about 1.2–1.5 Mt of tailings is transported to the TSF. The tailings are filtered to about 84% solids (approximately 80%–90% passing the No. 200 mesh size), desulfurized between 0.1% and 0.3% sulfur, including cyanide removal, before it is trucked to the TSF. The remaining tailings (approximately 0.17 Mt) and some of desulfurized tailings (approximately 0.7 Mt) are returned underground as cemented paste backfill.

The TSF footprint is approximately 80 ha and is designed to be built in four phases (i.e., cells), each with a surface area of about 20 ha. Each cell has about 5 years of storage capacity. The TSF footprint is lined with a geomembrane, which is anchored by a perimeter 5- to 10-m-high rock fill. Underlying the geomembrane, a 0.5-m-thick sand and gravel drainage blanket is provided, and the collected groundwater is conveyed to the collection pond, which is located at the west end of the TSF. The TSF will reach a maximum height of about 45 m and comprises four 10-m-high benches with a 4H:1V slope separated by a 10-m wide bench for an overall slope of 5H:1V. The tailings are placed and compacted in 0.5-m lifts. Phase 1 reached its maximum capacity in 2019, and the deposition of tailings in phase 2 began in the summer

Courtesy of Newmont

FIGURE 1 Tailings storage facility layout, June 2020

of 2018. Construction of phase 3A began in 2020 and was completed in 2021. The layout of the TSF is presented in Figure 1.

Sludge from the industrial water treatment process and dredging of the sedimentation pond is produced in smaller quantities. The sludge is filtered by using geotubes before it is transported to the TSF. About 20,000 t (metric tons) of sludge is annually disposed in the TSF. The underground mine also annually produces approximately 200,000–350,000 t of waste rock. This waste rock is stored in temporary stockpiles located in the industrial area and in the TSF. The waste rock (minus 0.3–0.5 m) is considered to be arsenic leachable.

In the initial closure plan, most of the waste rock, in the form of paste backfill, would be returned underground at the mining cessation to fill the galleries. However, the estimated amount of waste rock has significantly increased because of modifications in the mining methods and plans. As a result, the waste rock stockpile within the TSF footprint will not be removed prior to the phase 3 development, and the waste rock stockpile at the industrial site is at its maximum capacity.

The temporary waste rock pile in the phase 3 TSF area is shown in Figure 1B (in the background). The pile has an area of 11 ha and contains approximately 2.5 Mt (1.2 Mm^3) of waste rock. The waste rock pile is about 40 m high and consists of six 6-m-high benches separated by 10-m wide benches.

OPERATIONAL CHALLENGES WITH TAILINGS PLACEMENT

In the original design, no drainage measures were used within the TSF stack during deposition operations except for a minor allowance of mine rock for building access roads. During placement of the filtered tailings with the 50-t CAT trucks, travel on the tailings during the spring melt and rainy periods was difficult. It was decided to use waste rock to construct a network of access roads within the TSF to improve trafficability on the tailings surface. However, as the tailings stack developed, there were problems with site surface drainage that caused severe erosion of the slopes that resulted in erosion scars up to a maximum depth of 4 m. This also impacted the placement and compaction of the filtered tailings.

In addition, disposal of the sludge was originally located in an isolated section of phase 1. This was not practical because the TSF would lose considerable storage capacity as the filtered tailings could not be placed over the sludge material because of stability concerns. It was decided to place the waste rock on the advancing tailings front in separate locations to ensure stability of the tailings slope and then place the sludge material on top of the waste rock. The sludge could dry and consolidate during summer and fall before placement of the next lift of waste rock to encapsulate the sludge. Then, tailings placement could continue and cover the waste rock material.

WASTE ROCK CO-DEPOSITION WITH FILTERED TAILINGS

With the surplus of waste rock, it was decided to incorporate this material in the tailings stack by constructing a central drain toward the end of phase 1 and connecting drainage fingers (access roads). This would facilitate site surface drainage and also improve the tailings stack stability by lowering the phreatic level in the tailings. The central drain would be about 15 m wide and would be initiated 3 m above the compacted tailings overlying the geomembrane. This would prevent puncturing of the geomembrane. Details of the waste rock inclusions are shown schematically in Figure 2.

The introduction of waste rock into the tailings is considered as an operational improvement. Taking this into account, no transition zones and/or geotextiles between the tailings and waste rock were used, and in the long term, the voids in the waste rock will be eventually filled by tailings because they are not filter graded. As seen in Figure 2, site drainage is directed toward the interior of the TSF to avoid erosion of the exterior slopes. The central drains and rock-fill roads (fingers) also help to delineate the different areas for tailings deposition. In addition, the waste rock maintains a lower phreatic level in the tailings until the closure cover is implemented (cell 1 closure cover is planned for 2022). In addition, as the tailings stack is raised, the central drains are slightly located toward the interior, such that only one central drain exists in the final bench of the TSF.

The actual configuration of the waste rock inclusions is shown in Figure 3. Waste rock has been used for the access roads, central drains, and sludge disposal as represented by the various colors. Waste rock placement for phase 2 has been carried out in a more systematic manner. In phase 1, the central drain followed the alignment of the underlying transfer channel, as shown

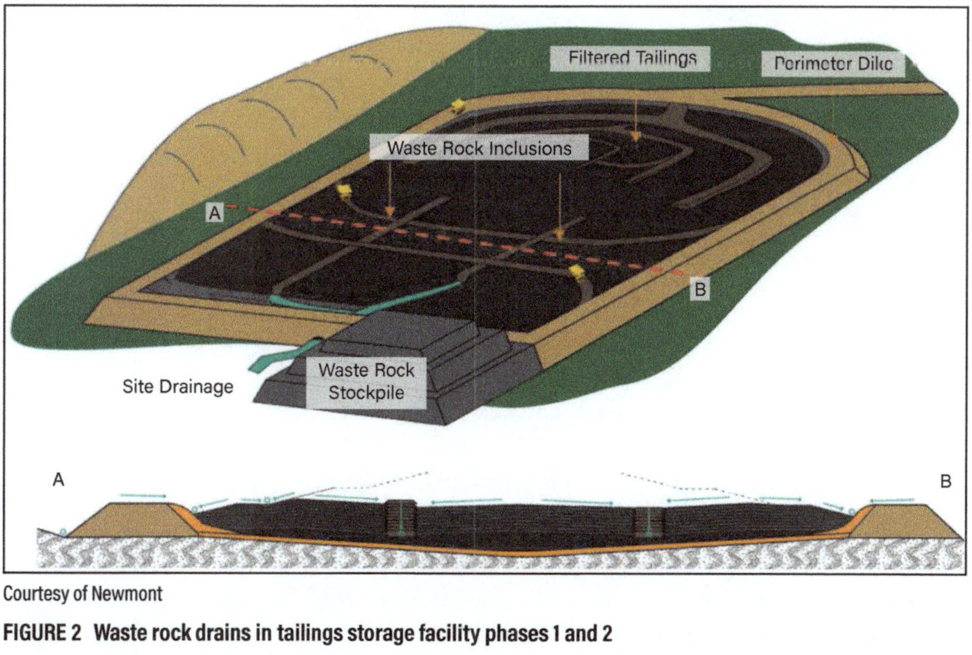

Courtesy of Newmont

FIGURE 2 Waste rock drains in tailings storage facility phases 1 and 2

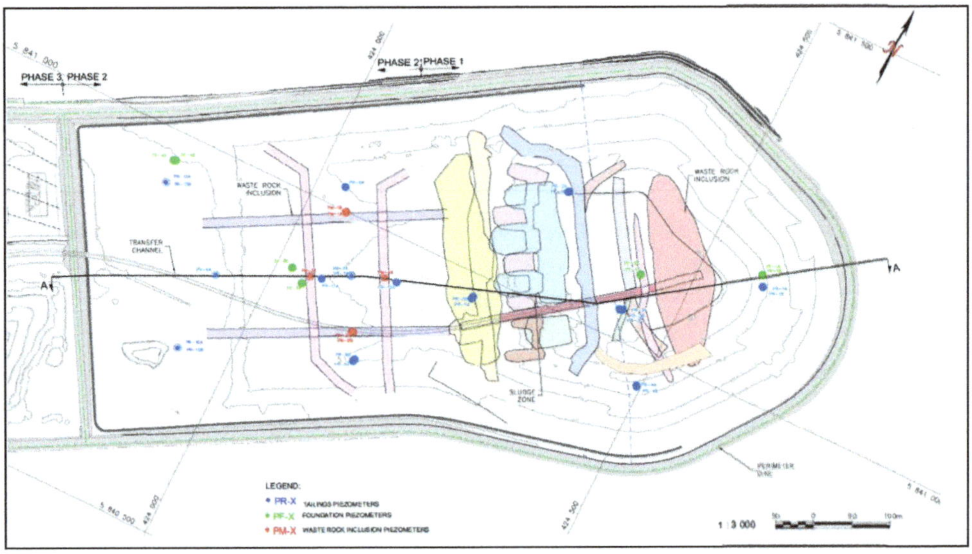

Courtesy of Newmont

FIGURE 3 Layout of waste rock placement and drains in tailings storage facility phases 1 and 2

in Figure 3. Furthermore, it was decided in the phase 2 development to build another central drain in the northern section of the stack because of the abundance of waste rock.

Monitoring the phreatic water level in the tailings is measured by vibrating wire piezometers (VWPs) by site personnel. To observe the performance of the waste rock drainage inclusions, additional VWPs have been installed adjacent to the central drains. Locations of the piezometers in the tailings stack and the basin foundation are shown in Figure 3.

Moving forward, it is planned to incorporate the existing waste rock stockpile into the tailings stack in combination with the construction of central drains and access road fingers. It is planned to relocate the waste rock from the industrial site stockpile to the TSF on a periodic basis until the end of closure to encapsulate the waste rock during TSF operations. Placement of the waste rock is not permitted in the outer shell (10-m-wide zone) and within the final few meters of the top surface, so as not to impact the closure design configuration.

BENEFITS OF THIS INTEGRATED WASTE CO-DEPOSITION STRATEGY

The co-deposition of waste rock and filtered tailings has significant benefits:

- Closure of the industrial site waste rock stockpile has been eliminated because the waste rock will be relocated to the TSF during site operations and the remaining waste rock will be returned underground at closure.
- Site drainage has been improved, thus ensuring the compaction and placement of the filtered tailings.
- The tailings stack stability has been improved by keeping the phreatic level low in the tailings until the closure cover is implemented.

One main disadvantage of incorporating waste rock is that the ultimate TSF storage capacity is reduced. However, with the current life of mine, there is sufficient capacity to jointly store the tailings and waste rock within the designed footprint. ■

Seismic Behavior of Chilean Downstream-Constructed Tailings Sand Dams

Luis Valenzuela
Institute of Mining Engineers

José Campaña
Arcadis

Chile is known as one of the most seismic countries in the world, responsible for more than 40% of the seismic energy globally liberated. The Valdivia earthquake (EQ) of 1960 with a moment magnitude (M_w) = 9.5 is the greatest EQ ever registered. In February 2010, the Maule EQ (M_w = 8.8) occurred, which is among the world's six largest EQs. In September 2015, the Illapel EQ (M_w = 8.4) occurred located about 300 km north of Santiago, Chile. Since 1906, about 23 EQs with M_w ≥7.5 have occurred. These EQs originated in the subduction zone of the Nazca Plate below the South American Plate, with EQ epicenters aligned along the Chilean coast, mainly north of parallel 40°S (Figure 1 and Table 1; Valenzuela et al. 2020).

Chile is also characterized as a narrow strip of land approximately 4,000 km long between the high Andes Mountains in the east and the Pacific Coast in the west, with an average width of about 200 km. Chile, which is the world's largest copper producer, presents significant mining activity, mostly north of parallel 35°S. In this scenario, all tailings facilities have been or will be subjected to one or more severe EQs in their life cycles.

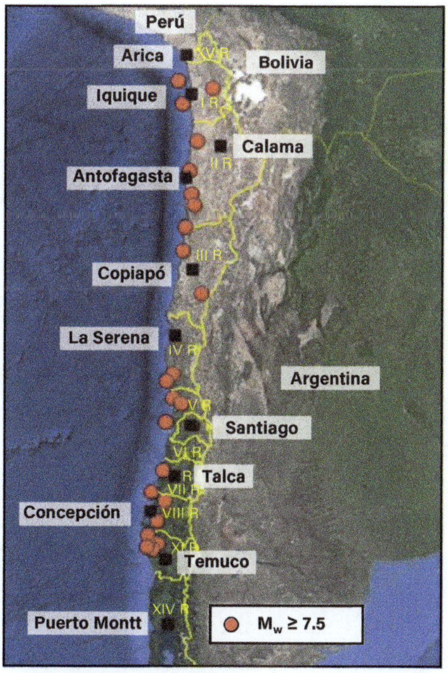

Source: Valenzuela et al. 2020

FIGURE 1 Earthquakes M_W ≥ 7.5 since 1850

Before the catastrophic failure of El Cobre tailings facility in 1965 as a result of the La Ligua EQ (M_W = 7.4) that provoked more than 200 casualties, most of the tailings embankments in Chile were built using the upstream method of construction. Construction of tailings embankments using the upstream method was prohibited in Chile as of 1970 through Decree 86 issued by the Ministry of Mining, unless a specific permit was issued by the national director of the agency for extraordinary conditions (Valenzuela 2016). However, it was not until 2007 that Decree 248

TABLE 1 Earthquakes of $M_w \geq 7.5$ since 1870

Date	Latitude	Longitude	Depth, km	M_w
September 16, 2015	-31.572	-71.674	22	8.3
April 3, 2014	-20.570	-70.493	22	7.7
April 1, 2014	-19.609	-70.769	25	8.2
February 27, 2010	-36.122	-72.898	23	8.8
November 14, 2007	-22.247	-69.890	40	7.7
June 13, 2005	-19.987	-69.197	115	7.8
July 30, 1995	-23.340	-70.294	46	8.0
March 5, 1987	-24.388	-70.161	62	7.6
March 3, 1985	-33.135	-71.871	33	8.0
May 10, 1975	-38.183	-73.232	6	7.7
July 9, 1971	-32.601	-71.076	60	7.8
December 28, 1966	-25.494	-70.550	25	7.7
May 22, 1960	-38.143	-73.407	25	9.5
May 22, 1960	-38.061	-73.039	25	7.8
May 21, 1960	-37.824	-73.353	25	8.1
May 6, 1953	-37.093	-72.866	66	7.5
April 6, 1943	-31.432	-71.475	35	8.1
January 25, 1939	-36.305	-72.315	35	7.8
July 13, 1936	-24.720	-70.110	35	7.5
December 1, 1928	-35.155	-72.105	35	7.7
November 11, 1922	-28.293	-69.852	70	8.5
December 4, 1918	-26.538	-70.608	40	7.8
August 17, 1906	-32.400	-71.400	35	8.2

Data from National Seismological Center, University of Chile

confirmed banning of tailings embankments constructed using the upstream method in any instance (Valenzuela 2016).

Since then, no new tailings facility has failed because of EQ loading; however, a few relatively small old inactive upstream embankments have failed, most with no major consequences, except for the Las Palmas flow failure in 2010 that killed four people (Valenzuela et al. 2020; Verdugo et al. 2012). This is a case in which continued upstream construction of an existing tailings embankment was performed, clearly against the principles of new legislation. New major tailings facilities built after 1965 were embankments using borrow materials, waste rock fill, and sand, with final heights varying from 50 to 237 m. Except for two cases of centerline construction, the remainder were constructed using the downstream method. No significant damage has occurred in any of these embankments despite 12 EQs with $M_w \geq 7.5$ occurring after 1965 (Valenzuela et al. 2020).

Six of these facilities with heights (measured vertically below dam crest) varying between 68 and 237 m (final height) are downstream-constructed tailings sand dams. They have quite satisfactorily performed, even under strong EQ loadings (Valenzuela et al. 2020). The Quillayes tailings facility is of special interest as it is presently the tallest sand dam in Chile with a final height of 198 m (Figure 2). The embankment was raised by downstream construction using hydraulic deposition of tailings sands containing no more than 18% fines (material passing ASTM 200 mesh). The inclined downstream deposition surfaces range from 3.5H:1V (horizontal: vertical) to 4H:1V and were compacted to 95% standard Proctor dry density by tandem bulldozers and smooth vibratory rollers. The starter embankment was 70 m high, constructed of compacted earth embankment fill. The dam foundation consists of fluvial deposits in the central area, colluvial and alluvial deposits in the right abutment, and intrusive rock in the left abutment (Campaña et al. 2015a).

Courtesy of L. Valenzuela © 2006

FIGURE 2 Quillayes tailings sand dam

Static and pseudo-static stability analyses using limit equilibrium methods were performed in 1997–1998. Seismic deformations were estimated by pseudo-dynamic analysis (Makdisi and Seed 1978) and FLAC2D and FLAC3D analyses (Itasca 1997). Numerical analysis (two- and three-dimensional) did not show evidence of failure by shear resistance or excess of deformation that could affect the global stability. The design EQ was the maximum credible earthquake (MCE) with M_w = 8.3, with epicenter 123 km from the dam and depth of 23 km, with a peak ground acceleration (PGA) of 0.37 *g* (horizontal) and 0.25 *g* (vertical) in hard soil, similar to EQ occurring in Illapel 125 km away from the dam in 1943 (Campaña et al. 2015a). A remarkably similar EQ (M_w = 8.4) occurred on September 16, 2015, at the same depth and near to the epicenter of the design EQ (Valenzuela et al. 2020). Unfortunately, no seismic records of the dam are available, but detailed post-EQ inspection of the dam did not detect significant deformations or instability, remarkably of lower magnitude than those indicated by the numerical analyses (Troncoso et al. 2017).

A similar behavior has been observed in other downstream sand tailings dams in Chile, all of them having been impacted by strong EQs ≥8.0 M_w (1985, 1995, 2010, 2015). The Tórtolas dam, located 60 km north of Santiago, is another good example. The dam has similar characteristics to the Quillayes dam and was subjected to the Maule EQ (M_w = 8.8) in 2010, with a horizontal PGA of 0.178 *g* registered downstream of the structure. Near the epicenter zone, a PGA of 0.65 *g* (horizontal) and 0.58 *g* (vertical) were registered. Dynamic modeling completed after the EQ showed deformations of 4–10 cm, consistent with field observations. Post EQ, piezometers installed upstream of the dam axis showed temporary increases in level up to 4.5 m. No significant variations in other instruments were detected (Valenzuela et al. 2020; Campaña et al. 2015b; González et al. 2019).

The main characteristics of Chilean downstream sand tailings dams are as follows (Valenzuela 2015; Verdugo et al. 2017):

- Specification of maximum fines content in the sand dam typically ranges from 15% to 20% to provide a free-draining shell with an ample contrast of permeability of the sand with respect to that of the tailings or slimes (Figure 3).
- A centralized cyclone station is fully instrumented to guarantee the quality of specified sand, mainly in relation to the maximum content of fines.

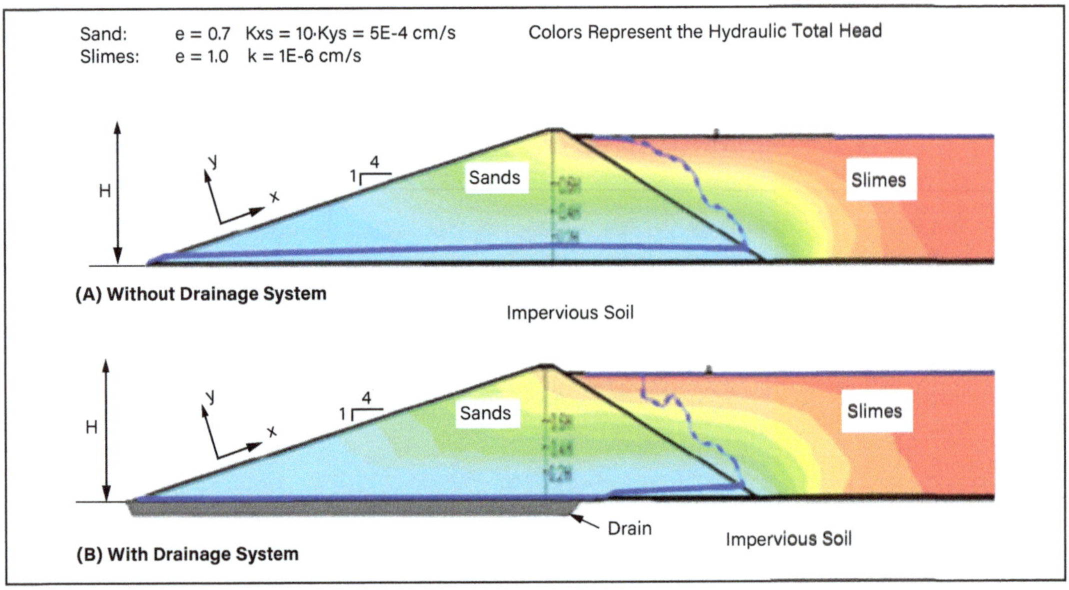

Source: Valenzuela 2015

FIGURE 3 Example of flow net in a typical cross section of a downstream sand tailings dam

- Sand is distributed from the crest of the dam over a gentle downstream slope to facilitate sand compaction with bulldozers (typically, slopes between 3.5H:1V and 4H:1V).
- Sand is compacted to a minimum of 95% standard Proctor dry density or equivalent to support dilatant behavior of the sand.

Extensive drainage, including basal and lateral finger drains, capable of near instantaneous removal of water during the construction/operation stage provide for low phreatic levels in the dam. Once operations are finished, the drains provide continued long-term capacity to support adequate closure conditions. ■

REFERENCES

Campaña, J., Alarcon, J.C., Valenzuela, L., et al. 2015a. The Quillayes sand tailings in Chile design and operation. In *Tailings and Mine Waste 2015: Proceedings of the International Conference on Tailings and Mine Waste.* Vancouver, Canada: University of British Columbia.

Campaña, J., Bard E., Cano, C., et al. 2015b. Seismic record in tailings deposit. [In Spanish]. Presented at the IX Chilean Congress of Geotechnics, Valdivia, Chile.

González, J., Torres, P., Campaña, J., et al. 2019. Verification of dynamic behavior modeling of Las Tórtolas main dam during 27F Chile earthquake. In *Tailings 2019 Online Proceedings*. Santiago, Chile: Gecamin. www.gecaminpublications.com/tailings-2019. Accessed February 2021.

Itasca. 1997. *FLAC2D* ™and *FLAC3D* ™. Minneapolis: Itasca Consulting Group, Inc.

Makdisi, F.I., and Seed, H.B. 1978. Simplified procedure for estimating dam and embankment earthquake-induced deformations. *Journal of the Geotechnical Engineering Division* 104(7):849–867.

National Seismological Center. 1906–2015: Earthquakes of moment of magnitude greater than or equal to 7.5 since 1870. Santiago, Chile: University of Chile.

Troncoso, J.H., Verdugo, R., and Valenzuela, L. 2017. Seismic performance of tailings sand dams in Chile. In *Proceedings of the 16th World Conference on Earthquake Engineering: Resilience: The New Challenge in Earthquake Engineering*. Santiago, Chile: Chilean Association on Seismology and Earthquake Engineering.

Valenzuela, L. 2015. Tailings dams and hydraulic fills: The 2015 Casagrande lecture. In *Advances in Soil Mechanics and Geotechnical Engineering.* Vol. 5: Geotechnical Synergy in Buenos Aires. Edited by A.O. Sfriso, D. Manzanal, and R.J. Rocca. Amsterdam, The Netherlands: IOS Press.

Valenzuela, L. 2016. Design, construction, operation and the effect of fines content and permeability on the seismic performance of tailings sand dams in Chile. *Revista Obras y Proyectos* 19:6–22.

Valenzuela, L., Verdugo, R., Campaña, J., et al. 2020. Safety performance of dams in Chile highly seismic environment. Presented at the 4th International Dam World Conference [webinar], Lisbon, Portugal.

Verdugo, R., Sitar, N., Frost, J.D., et al. 2012. Seismic performance of earthen structures during the February 2010 Maule, Chile, earthquake: Dams, levees, tailings dams, and retaining walls. *Earthquake Spectra* 28(S1):S75–S96.

Verdugo, R., Campaña, J., and Valenzuela, L. 2017. Chilean practice on large tailings dams constructed on high seismic environment. In *16th World Conference on Earthquake Engineering Newsletter No. 4.* Santiago, Chile: Chilean Association on Seismology and Earthquake Engineering.

Seismic Deformation Analysis of Tailings Impoundment No. 2, Mine Niobec, Canada

Michael James
Agnico Eagle Mines

Typically, upstream-raised embankments are constructed such that the raises are partially supported by the retained tailings, which are often loose (contractive) with a high phreatic surface near the downstream slope. This configuration renders them particularly vulnerable to ruptures induced by the liquefaction of the retained tailings. This case study briefly summarizes the design and construction of an upstream-raised embankment without this vulnerability.

Mine Niobec is located in the Saguenay region of the province of Quebec (Canada), about 200 km north of Quebec City. Slurry tailings from an underground niobium mine are stored in impoundments on site. The tailings are nonplastic and vary from colloidal to fine sand (0.2 mm) in grain size. The mine lies in a zone of moderate seismicity that produced the moment magnitude (M_w) 5.9 Saguenay earthquake and is capable of producing events with magnitudes as high as 7.5. Tailings impoundment No. 2 at the site was designed and constructed in the 2000s with the potential for seismic liquefaction of the tailings duly considered.

The impoundment is rectangular with plan dimensions of 650 m by 1,250 m and a maximum height of 30 m. To allow construction in the upstream direction while avoiding the vulnerability usually associated with this concept, the downstream slope consists of an internally drained, compacted shell of the coarser fraction of the tailings. Two starter dikes of compacted coarse tailings, 60 m apart, were constructed around the perimeter of the impoundment. A filter-protected drainage system of perforated pipes and French drains at 23 m intervals extends between the starter dikes and passes through the exterior starter dike, conveying drainage to exterior collection ditches. From May to December, coarse tailings were placed between the starter dikes and compacted using the cell method, which consists of depositing the slurry in a small space confined by low berms and compacting it with a loaded bulldozer. The coarse fraction of the tailings is sufficiently permeable for adequate compaction (a density index, I_D, of at least 75%). The compacted shell of coarse tailings was raised in the upstream direction with a 4:1 (H:V) external slope and a 2:1 internal slope; the shell decreases in width with height. From May to December, the finer fraction of the tailings was deposited as slurry upstream of the shell and during the remaining months, the total tailings were deposited as slurry. A typical cross section of the impoundment is shown in Figure 1. Construction monitoring was accomplished by nuclear density testing, and seismic cone penetration testing was used to verify the characteristics of the compacted tailings with depth as the impoundment was raised (Lemieux et al. 2011).

The seismic stability of the tailings facility was verified using nonlinear dynamic analysis (finite difference method). A two-dimensional geometric model of the tailings facility was

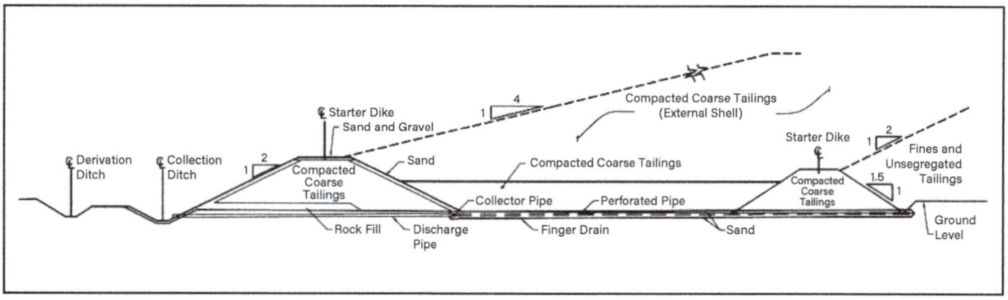

Source: Lemieux et al. 2011

FIGURE 1 Typical cross section of tailings impoundment No. 2 (not to scale)

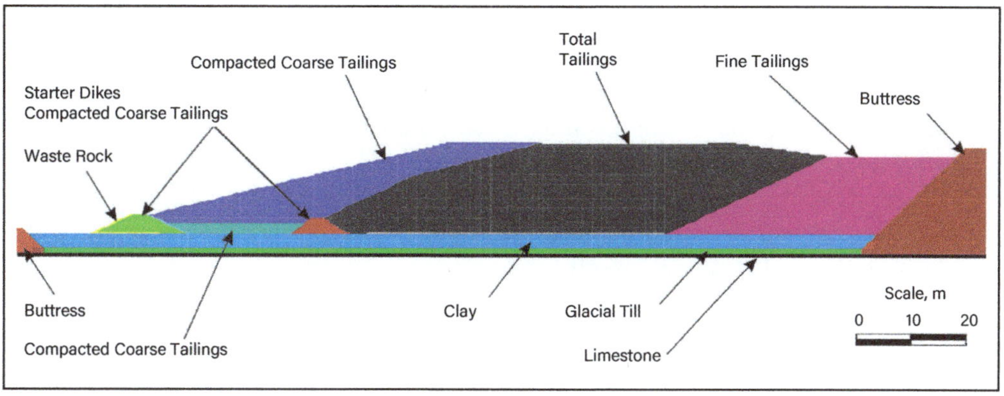

Source: James and Leahy 2011

FIGURE 2 Finite difference model of tailings impoundment No. 2

created (Figure 2). An elastic model was used for bedrock (competent limestone), the Mohr-Coulomb model was used for overburden (clay and till), and the UBCSAND model (Byrne et al. 2004) was used for tailings (compacted and slurry-deposited). A series of ground motions compatible with the design earthquake ground motion (a 1,000-year event) was used to simulate the occurrence of earthquakes. One of these ground motions was a factored signal from the 1988 Saguenay event (S-16L), which has factored peak ground acceleration of 0.2g and an Arias intensity of 0.25 m/s. Interpretation of these results, which were typical, indicated moderate generation of excess pore-water pressure in the retained tailings (maximum excess pore-water pressure ratio, r_u, of 0.3) and no deformation of the compacted shell of engineering significance. Analysis of the conditions at the end of shaking (excess pore-water pressure development, shear strains, nodal velocities), indicated stability and that post-shaking simulations were not necessary (James and Leahy 2011). ■

REFERENCES

Byrne, P.M., Park, S., Beaty, M., et al. 2004. Numerical modeling of liquefaction and comparison with centrifuge tests. *Canadian Geotechnical Journal* 41:193–211.

James, M., and Leahy, D. 2011. A case study: The seismic stability of an upstream-raised tailings impoundment (Part II). Presented at the 2011 Pan-Am CGS Geotechnical Conference, Toronto.

Lemieux, N., Leahy, D., St-Laurent, J.F., et al. 2011. A case study: The seismic stability of an upstream-raised tailings impoundment (Part I). Presented at the 2011 Pan-Am CGS Geotechnical Conference, Toronto.

Tailings Storage Facility Developed Within a Heap Leach Facility, Yanacocha Mine, Peru

Anamaria Rios Pando
Minera Yanacocha S.R.L.

Efficient use of space is often a priority for mining operations, and construction of a tailings storage facility (TSF) within a heap leach facility (HLF) is an example of a design driven by this priority. The Yanacocha operation is an open pit gold mine located in mountainous terrain in highly seismic northern Peru. At Yanacocha, implementation of a mill later in the project life cycle to process higher-grade ore from mine pits resulted in the need for tailings storage. Given topographic constraints and permit boundaries, the La Quinua (LQ) TSF was constructed entirely within one of the four site HLFs (LQHLF).

The LQHLF has been in operation since the mid-1990s, with the first phase of the TSF (termed the *LQ South TSF*) commissioned in 2008. The LQ South TSF was progressively developed within the LQHLF as the HLF was loaded with ore using large fleet equipment (i.e., CAT 793 haul trucks). The LQ South TSF was constructed in two phases: (1) the starter facility provided capacity for approximately 45 Mt (million metric tons) of tailings; and (2) a 12-m-high centerline raise was constructed to provide an additional 31 Mt of capacity. A second cell, fully excavated into the surface of the LQHLF with a depth of 45–80 m, and founded on leached ore, was added in 2015 (termed the *LQ North Phase 1 TSF*) to receive tailings once capacity in the LQ South TSF was depleted. The LQ North Phase 1 TSF was constructed with a capacity to store 18 Mt of tailings. Figure 1 is a view of the LQ HLF with deposition ongoing

Source: Prado et al. 2016

FIGURE 1 View of the La Quinua (LQ) South tailings storage facility (TSF) (left) and the LQ North Phase 1 TSF (right) during construction

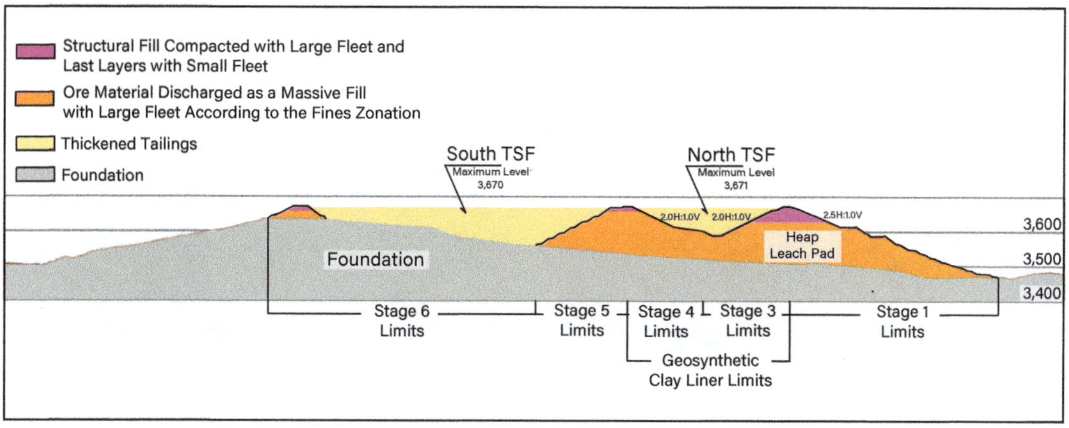

FIGURE 2 Cross section A through the La Quinua (LQ) heap leach facility showing the LQ South tailings storage facility (TSF) (left) and the LQ North TSF (right)

in the LQ South TSF and the LQ North Phase 1 TSF under construction. A cross section through the facility is included in Figure 2.

The mill tailings are subjected to cyanide as part of the gold recovery process and thickened to a high-density slurry (nominally 62% by weight) prior to placement within the TSF. Tailings are produced at a nominal rate of 16,500 t/d (metric tons per day). A liner system is installed within each of the TSF cells to limit seepage and tailings migration into the surrounding heap with a partial liner installed in the LQ South TSF and a full liner within the LQ North Phase 1 TSF.

Key design considerations included:

- Stability and deformation potential of the leached ore slopes under static and seismic loading conditions
- Ability to efficiently remove water from the lined tailings impoundments
- Heterogeneous nature of the leached ore including potential for development of perched phreatic conditions

To provide reliable inputs to geotechnical engineering analyses, investigations completed during the design process included seismic cone penetration tests with pore pressure dissipation tests, sonic drilling with standard penetration tests, and an extensive laboratory testing program on leached ore and tailings materials. Analyses completed to support the design included liquefaction assessment of the leached ore; slope stability analyses under static and seismic loading; evaluation of permanent deformation under seismic loading; and seepage analyses of the surrounding leached ore.

The combined facility comprised of the LQHLF, LQ South TSF, and LQ North Phase 1 TSF (Figure 3) has consistently performed within its design expectations. Plans are underway to further expand the tailings storage capacity within the LQHLF, including expansion of the LQ South TSF and construction of an additional cell (i.e., LQ North Phase 2 TSF). For more information on this case study, readers are referred to Prado et al. (2016). ■

Courtesy of Newmont

FIGURE 3 La Quinua (LQ) North tailings storage facility (TSF) (right) and LQ South TSF (left), January 2020

REFERENCE

Prado, F.E.G., Pando, A.R., Kerr, T.F., et al. 2016. Tailing storage within a heap leach facility—an update on the success at Yanacocha. In *Proceedings of Heap Leach Mining Solutions 2016.* Vancouver, British Columbia, Canada: InfoMine.

Use of Wick Drains to Stabilize a Filtered Tailings Facility

Bryan Ulrich
Bryan Ulrich LLC

A filtered tailings facility in Latin America was in the early stages of operation when bulging was noted to have occurred along its lower slopes. The facility was a cross-valley–type structure. It included a rock starter berm; and a contact water pond, which accepted surface water runoff; and water from the underdrain system. Clean stormwater was diverted around the facility by diversion channels. Filtered tailings were delivered to the facility by conveyor, after which it was worked into place by backhoes, trucks, and dozers.

The operators were preparing for the first "step-in," or bench, which would be receiving rock cladding to protect the facility against erosion. Dozers had leveled this first bench in preparation of rock placement. The following day, the operators returned to the site, and noted that the area that had been leveled was no longer level, and the lower slope of the facility had noticeably bulged.

Following coordination with the owners, in situ testing was organized, including a number of cone penetration tests and the installation of drive-point vibrating wire piezometers. This information was used to update the slope stability analyses, including for the current conditions as well as for the ultimate facility. An assessment of the data indicated that there were areas of considerable pore pressures within the filtered tailings deposit, some of which were in excess of hydrostatic pore pressures. The magnitude and distribution of pore pressures were found to be unacceptable for the slope stability of the facility, and thus remedial measures were sought. Such measures considered included buttressing, slope flattening, horizontal drain

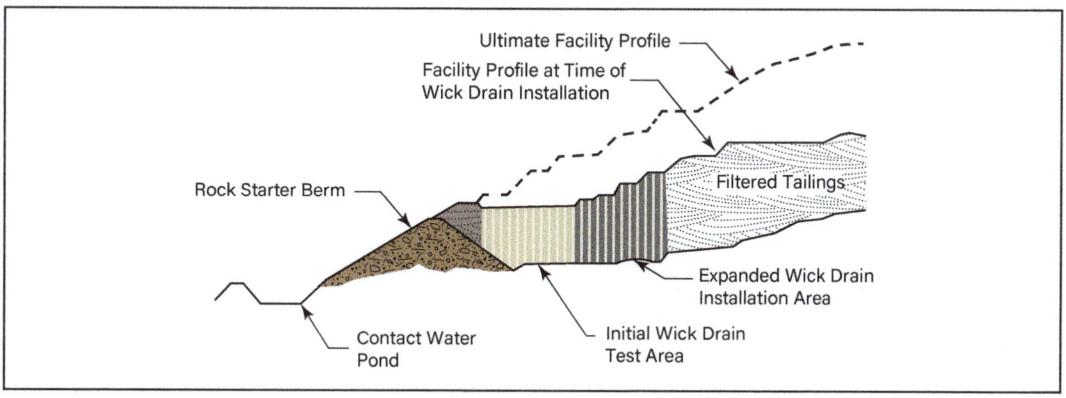

Courtesy of Wismut GmbH

FIGURE 1 Profile through facility showing areas with wick drains installed

placement, and electrical methods to remove water. The decision was made to install vertical wick drains. Initially, a test installation was commissioned using a specialty contractor to allow an assessment to be made of the actual performance of the wick drains. The location of the test installation is shown in Figure 1. Additional drive-point vibrating wire piezometers were also installed, and the decay of the pore pressure was monitored. The performance was deemed a success, and an expanded area (shown in the figure) was identified for the larger-scale wick drain installation area, based on the seepage and slope stability analyses. This installation largely was performed on benches that were above the lower step-in. The top of each wick drain was tied to a surface network of wick drain material, all of which was gravity fed into a stormwater decant structure, which was noted to run with a gentle but steady stream of water.

Data from the vibrating wire piezometers were monitored closely and advice provided to the facility operators when pore pressures starated to become elevated so that placement of the filtered tailings could be carried out in a safe manner. Through close cooperation, the facility was carefully operated until its eventual closure. ■

Cone Penetration Test Results Reflect Operational Improvements, Peñsaquito Mine, Mexico

E.J. García Romero
Newmont Peñasquito

V.V. Sosa Dávila
Newmont Peñasquito

The tailings storage facility (TSF) at Newmont's Peñasquito mine is a four-sided structure with an 11-km perimeter and a maximum height of approximately 93 m (Figure 1). The TSF receives approximately 120,000 t/d (metric tons per day) of tailings and has been in operation since 2009. The TSF is raised using the centerline method on three sides (north, west, and south) and the downstream method near the internal reclaim pond (on the east). The TSF is designed to be raised to a maximum height of 120 m with an ultimate storage capacity of 808 Mt (million metric tons).

Tailings deposition is carefully managed for successful operation of the TSF. Tailings must be deposited so that the beach rises relatively evenly around the facility. Even deposition is required for successful management of the reclaim pond location and its water quality and for maintaining a minimum 200-m exposed beach length (north, west, and south embankments only). Additionally, continuous construction of the TSF is necessary to maintain the required 2-m minimum freeboard.

Courtesy of Newmont

FIGURE 1 Tailings storage facility at the Peñasquito mine, Mexico (looking east)

Periodic cone penetration test (CPT) campaigns are performed to support ongoing characterization of the in-place tailings and phreatic conditions, as well as facilitate installation of additional monitoring instrumentation. A CPT campaign was executed in 2019 with the objectives of investigating tailings geotechnical properties, changes in tailings characteristics over time, and installation of vibrating wire piezometers (VWPs). A total of 22 non-seismic and 38 seismic CPT soundings were performed, as well as the installation of 67 VWPs.

CPT soundings were performed on the tailings beach, with soundings completed as far as 250 m upstream from the centerline of the dam. The soundings were located along 12 alignments oriented perpendicular to the dam crest, with CPTs advanced at approximately 50-m spacing along the alignments (i.e., generally five soundings performed along each alignment).

Data analysis was carried out that evaluated ways to segregate and summarize the CPT data into zones proximal (<100 m) and distal (>100 m) from the dam centerline. Cone tip resistance (q_c, q_t) and sleeve friction (f_s) data from the shallower intervals in the proximal soundings represented an interbedded sand to sandy-silt sequence in thick lithological intervals. The distal results showed sand and silt mixtures alternating in thin to very thin patterns as correlated to the index classification (*Ic*) or soil behavior type by Robertson and Cabal (2014) in Golder (2020).

Comparison of results of a previous 2015 CPT campaign to the 2019 campaign with respect to the *Ic* allowed for the evaluation of the effectiveness of the ongoing operational tailings management procedures. Along with *Ic* and its correlation with position in the impoundment and distance from the perimeter tailings spigots, hydrogeologic conditions are represented by having lower pore pressures on soundings closer to the dam centerline.

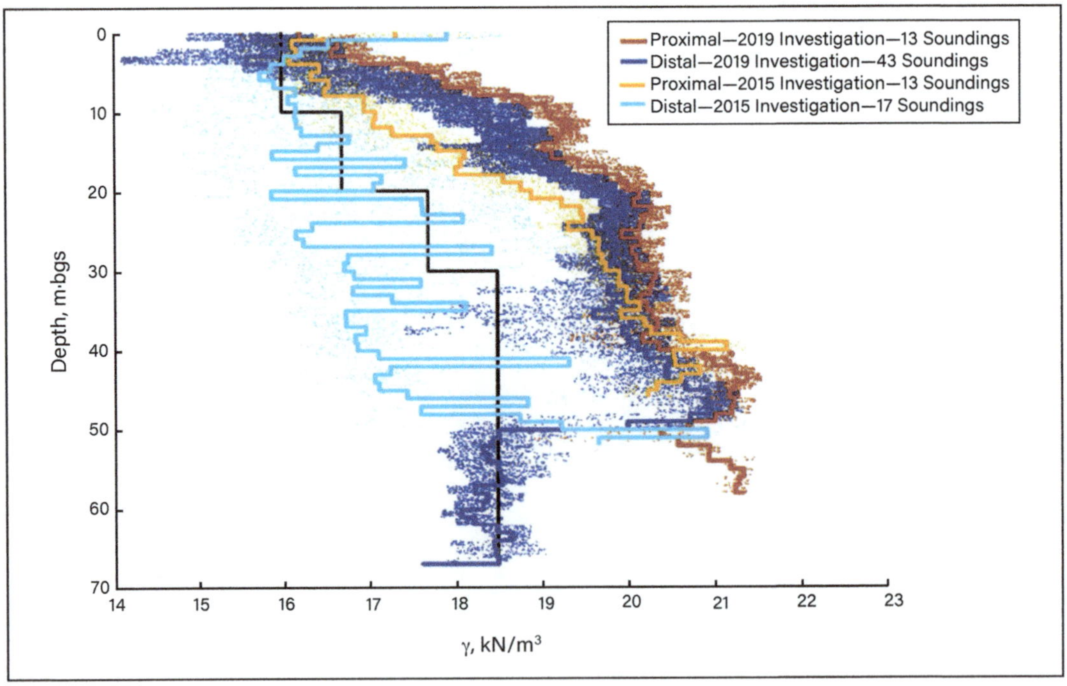

Source: Golder 2020

FIGURE 2 Comparison of 2015 to 2019 results in density profiles with respect to depth below the tailings surface

In-place tailings densities interpreted from the CPT data in 2019 were within an average of 17 to 21.5 kN/m^3, increasing progressively with depth in the tailings column. The 2015 data indicated erratically distributed in-place densities between 16.5 and 19.5 kN/m^3. The density trends with distance into the impoundment for the 2015 and 2019 campaigns are shown in Figure 2.

Pore pressure dissipation tests were performed during the CPT soundings to improve understanding of the phreatic conditions within the facility during the 2015 and 2019 CPT campaigns. The phreatic surface evaluated from the 2019 CPT campaign was significantly improved, extending farther upstream of the dam as compared to the phreatic surface obtained in 2015 (Figure 3). The distance at which the tailings become saturated is therefore farther from the dam crest. Specifically, in 2015, tailings were fully saturated to the surface approximately 100 m upstream from the centerline of the dam; while in 2019, saturation was generally reached >250 m into the impoundment. In 2015, the phreatic surface was located within the cyclone sand section of the dam at an elevation above the starter dam. As of 2019, the same VWPs that previously showed water within the dam centerline are now dry, and the phreatic surface exits below the starter dam within the foundation drains.

Periodic CPT campaigns carried out in the Peñasquito TSF provide valuable information on characteristics of the tailings over time. By performing the CPT programs, significant additional data are obtained that are incorporated into three-dimensional models of the TSF,

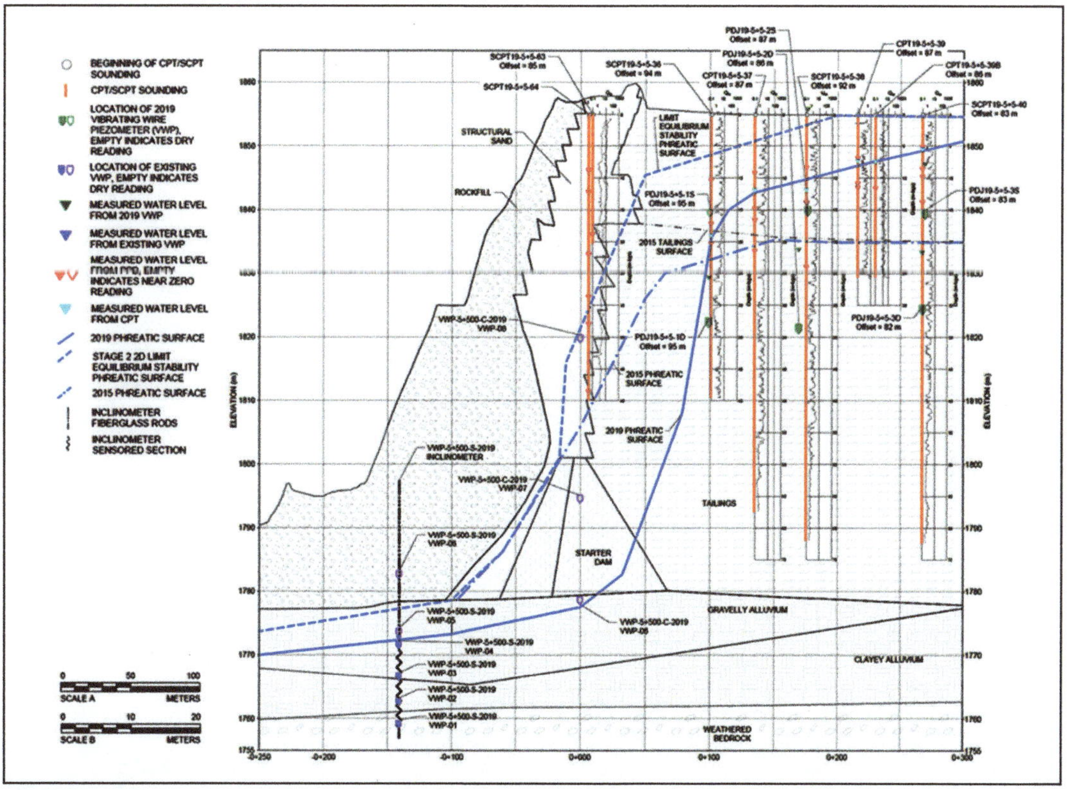

Source: Golder 2020

FIGURE 3 Results of 2019 cone penetration test campaign, detail of geotechnical instrumentation, and comparison of phreatic surfaces obtained in 2015 and 2019

stability analyses, and other geotechnical analyses; support optimization of design and management of the TSF; and provide insight on facility performance.

Tailings management practices are continuously improving at the site, with significant improvements incorporated during 2015–2019 as observed by the CPT data. This included improved beach management practices and reducing water stored within the internal pond. Beach management is a daily team integrated activity that provides a link to planners, surveyors, and operators together with construction quality assurance performed by the engineer of record. Further improvements are underway, including commissioning of an external pond system to further reduce the volume of water stored on the facility. For more information on this case study, readers are referred to García and Sosa Dávila (2020). ■

REFERENCES

García, E.J., and Sosa Dávila, V.V. 2020. Recent geotechnical monitoring results reflect operational improvements at the Newmont Peñasquito. In *Tailings and Mine Waste 2020: Proceedings of the International Conference on Tailings and Mine Waste*. Vancouver, Canada: University of British Columbia.

Golder. 2020. *2019 CPT Tailings Geotechnical Factual Report*. Vancouver, BC, Canada: Golder Associates.

Robertson, P.K., and Cabal (Robertson), K.L. 2014. *Guide to Cone Penetration Testing for Geotechnical Engineering*, 6th ed. Signal Hill, CA: Gregg Drilling and Testing.

Geotechnical Studies to Improve Understanding of Material Parameters and Tailings Storage Facility Performance and Address Changes in Stability Requirements, Granny Smith Gold Mine, Western Australia

Johan Boshoff
Gold Fields

The engineer of record (EoR) undertook a stability assessment of the tailings storage facilities (TSFs) at Gold Fields' Granny Smith mine (GSM). The stability assessment included limit equilibrium analyses and dynamic deformation modeling. The project also included supplementary studies in support of the stability analyses: a site-specific seismic hazard assessment and the development of a geologic model.

BACKGROUND

The GSM TSF is located 20 km south of the town of Laverton, Western Australia. The facility is raised using the upstream method and comprises three cells. The approximate maximum embankment heights of cell 1 and cell 2 are 33 m, while cell 3 is 17 m high. A plan view of the GSM TSF is provided in Figure 1.

The TSF was designed before the Australian National Committee on Large Dams (ANCOLD) established its guidelines on tailings embankments, with a minimum allowable factor of safety for static conditions of 1.3.

Previous investigations of cell 1 identified horizontal layering, resulting from the early deposition of primary and oxide tailings. The oxide layers were located below and behind the original starter embankment and covered approximately 50% of the TSF floor area.

The EoR estimated that the oxide tailings would likely exhibit brittle behavior with very low residual undrained shear strength ratios. As a result, preliminary stability analyses yielded factors of safety below 1.5 under undrained static loading conditions. In response to this lower factor of safety, deposition into cell 1 was suspended by Gold Fields, with deposition diverted to cell 3. Preliminary designs for a toe buttress were also commissioned.

OBJECTIVE OF ADDITIONAL STUDIES AND INVESTIGATIONS

The objective of the project was to confirm the post-seismic stability of each cell of the TSF and optimize the geometry of the toe buttress to address (as a minimum) the ANCOLD factor of safety requirement of 1.1. Dynamic deformation modeling will be used to understand the

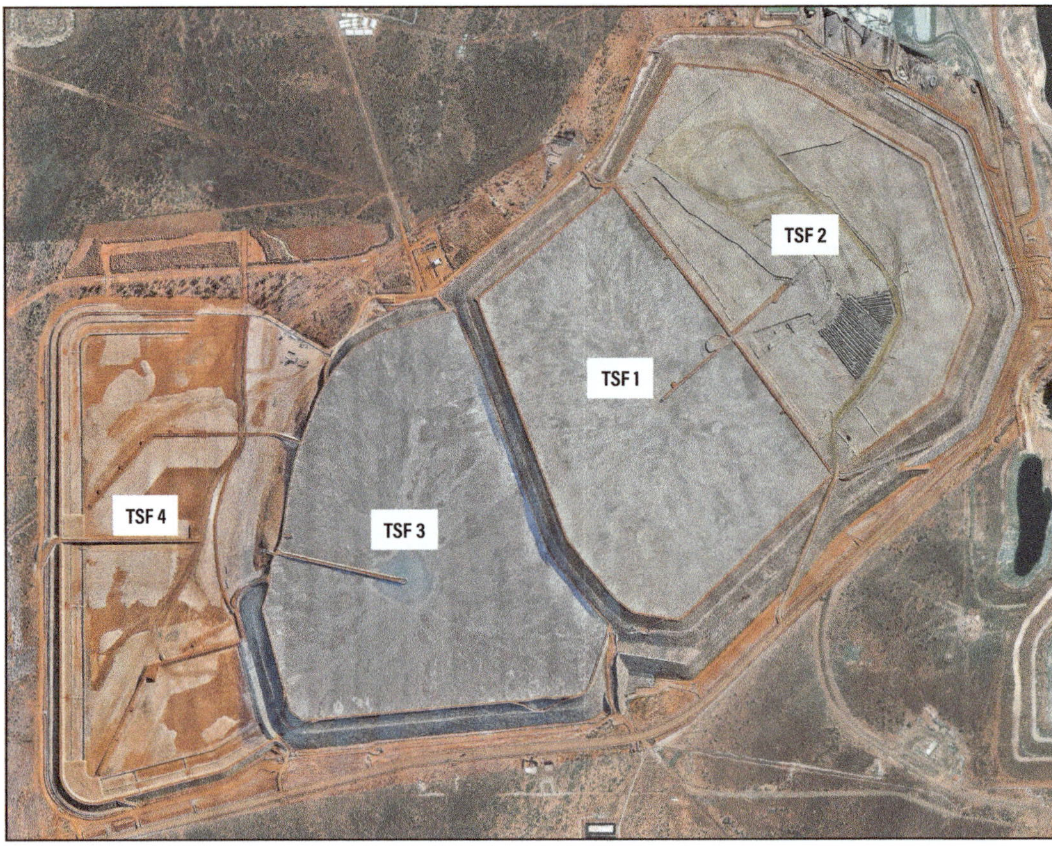

Courtesy of Gold Fields

FIGURE 1 Plan view of Granny Smith mine tailings storage facility

stress states, bias, and principal stress directions in the tailings to predict the deformation of the TSF under the seismic design load and assess the serviceability of the TSF cell.

GEOTECHNICAL INVESTIGATION

Geotechnical investigations were undertaken to assess the characteristics of the oxide tailings deposited near the starter embankments during the early stages of embankment development to improve the understanding of the state of the tailings deposited within cells 1, 2, and 3; establish geotechnical parameters to be used for the stability assessment of the TSF cells for evaluation, design, and extent of the toe buttress along cells 1 and 2; identify the extent and nature of foundation soils beneath the TSF embankments; and assess whether the foundation is critical to the stability of the TSF and buttress design.

The investigation confirmed that a very loose layer ($\Psi > 0.15$, likely highly brittle) could be inferred at the base of cell 1, potentially extending to the starter embankment (Figure 2). A second very loose layer could be inferred within cell 1, at elevations above the starter embankment. P_{85} ψ (85th percentile [P_{85}] characteristic state parameter value) within the very loose layers reaches values greater than 0.15, which could indicate the presence of pluviated oxide tailings resulting from maintaining a large supernatant pond near the embankment over some time. Several thin, perched saturated zones could be inferred within the areas where the cone

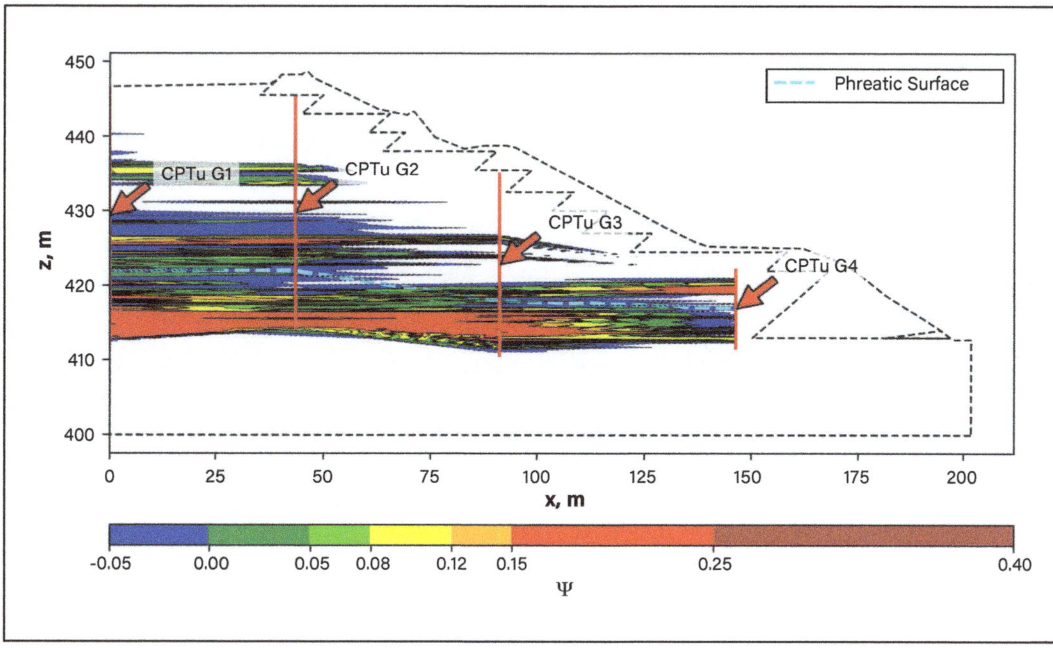

Source: Golder 2020

FIGURE 2 Critical section through cell 1 indicating layering

penetration test with pore water pressure measurement (CPTu) probing was undertaken, which generally appeared to coincide with the presence of loose tailings layers, potentially of high fines content.

THREE-DIMENSIONAL GEOLOGIC MODEL

A conceptual three-dimensional (3-D) ground model of the TSFs was constructed using the commercial 3-D visualization and modeling package Leapfrog Works (Seequent, n.d.). The model provides a 3-D interpretation of the stratigraphic units near the existing and proposed future TSFs (Figure 3).

The objectives of the 3-D geologic model were to

- Provide conceptualization of the geologic setting near the existing and proposed future TSFs
- Provide information on subsurface conditions
- Present new and historical 3-D data
- Provide geologic cross sections for numerical analysis of embankment design

LABORATORY TESTING

Regarding the laboratory testing campaign, the scope of work encompassed the following:

- Conducted classification and advanced laboratory testing on selected foundation, starter embankment, and tailings materials. This involved comprehensive analysis to determine the physical and mechanical properties of these materials.
- Interpreted the geotechnical behaviors of the materials, focusing on both monotonic and cyclic resistance. The laboratory test results were analyzed to gain insights into the

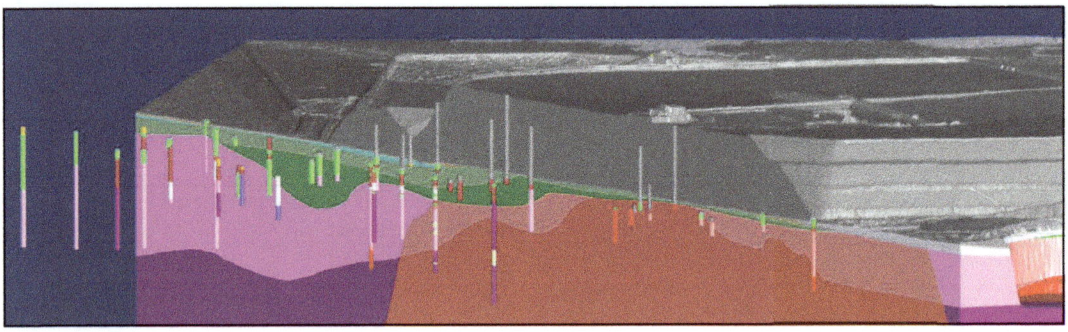

Source: Golder 2020

FIGURE 3 Extract from the 3-D geologic model, showing borehole geology relative to 3-D geologic volumes

response of the materials under different loading conditions, including static and cyclic loading.

- Estimated material parameters for slope stability and deformation analyses. This involved utilizing the limit equilibrium method (LEM) and performance-based numerical deformation analysis. By determining the material parameters, such as shear strength, stiffness, and deformability, it was possible to perform detailed stability and deformation analyses to assess the performance of the TSF under various loading scenarios.

SLOPE STABILITY IMPROVEMENT

After completing the geotechnical investigations and confirming in situ conditions, modeling was undertaken to represent the highest potential risk of instability based on geometry, construction, and loading conditions.

Static limit equilibrium assessments of the critical sections yielded factors of safety of ~1.5. The post-seismic stability analyses, based on very low strength ratios (~0.1) and representing brittle behavior, yielded factors of safety ~1. With the addition of a buttress, the post-seismic factors of safety increased to ~1.1. In addition, the buttress geometry was optimized based on the final laboratory testing results and deformation models.

Following completion of the proposed toe buttress, cell 1 will be available for contingency storage. An aerial image of the completed toe buttress can be seen in Figure 4.

Courtesy of Gold Fields

FIGURE 4 Cell 1 toe buttress

INSTRUMENTATION

Instrumentation was installed below the buttress footprint to monitor pore pressures in the foundation during the construction of the buttress. Other monitoring technologies that were installed included real-time accelerometers, seismic geophones, fiber-optic geophones, and radar.

LESSONS LEARNED

- Develop a detailed commissioning plan promoting thin-layer deposition and pond control during even the early days of embankment development.
- Identify clayey (problem) materials in the mining area and anticipate the impact these materials could have on the performance of the TSF.
- Recognize the importance of material characterization and understanding for upstream raised TSFs.
- Brittle materials are a special case and require conservative design.
- Changes in operational practice and regulation are likely to impact future designs and guidelines (i.e., ANCOLD guidance changes [ANCOLD 2019]).
- Recognize the value of thorough independent reviews. ■

REFERENCES

ANCOLD (Australian National Committee on Large Dams). 2019. *Guidelines on Tailings Dams: Planning, Design, Construction, Operation and Closure*, revision 1. Hobart, Tasmania, Australia: ANCOLD.

Golder. 2020. Preliminary limit equilibrium stability assessment and indicative buttress requirements. Technical Memorandum No. 20137760-007-M-Rev1, November 17. Toronto, Ontario, Canada: Golder Associates Ltd.

Seequent. n.d. *Leapfrog Works* [software application]. Vancouver, British Columbia, Canada: Seequent Limited.

Tailings Deposition in Open Pits at Cameco's Key Lake and Rabbit Lake In-Pit Facilities, Northern Saskatchewan, Canada

Barry Esford
Cameco Corporation

Scott Donald
WSP

Cameco Corporation operates uranium mills and tailings management facilities at its Key Lake and Rabbit Lake operations in Northern Saskatchewan, Canada. Development at both mine sites began in the 1970s, with initial tailings deposition occurring within above-ground impoundments, which was common for these facilities at the time. As open pits became available at each site, and mining continued at other ore bodies in the area, tailings management shifted to in-pit facilities to meet stakeholder expectations and improve the long-term security and environmental performance of tailings management at these facilities.

The in-pit facility at Rabbit Lake (receiving tailings since 1984; Figure 1), was one of the first (if not *the* first) engineered in-pit tailings facilities in the world and adopted a full

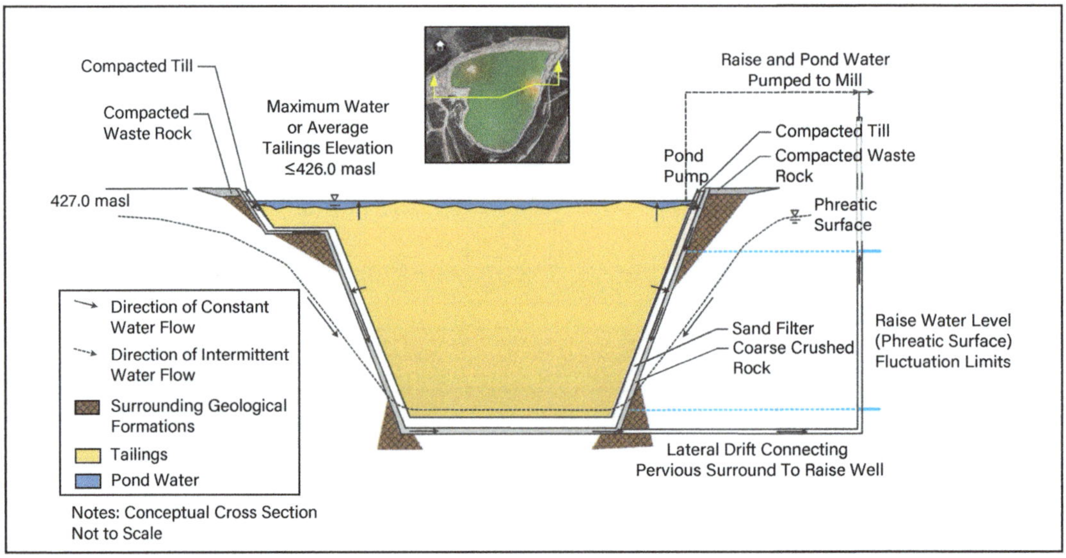

Courtesy of Cameco Corporation

FIGURE 1 Cameco's Rabbit Lake operation in-pit tailings storage facility

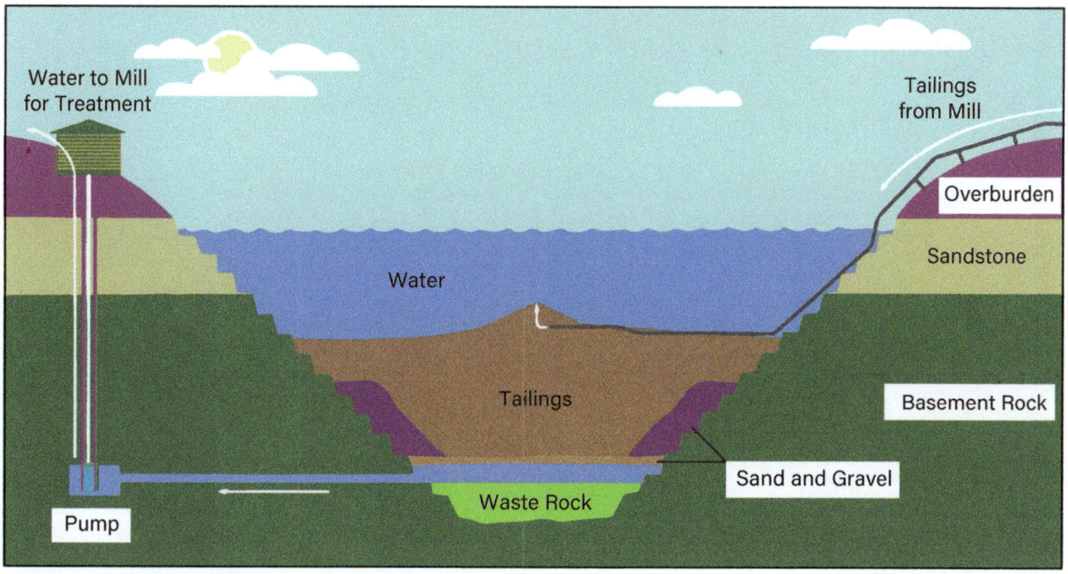

Courtesy of Cameco Corporation

FIGURE 2 Cameco's Key Lake operation in-pit tailings storage facility

pervious surround design concept with subaerial deposition. Under this concept, permeable sand and coarse rock are placed at the base of the pit, around its perimeter, as well as overlying the tailings at closure to promote regional groundwater flow bypassing the lower-permeability tailings. In contrast, at Key Lake, the in-pit facility (receiving tailings since 1996) adopted a partial pervious surround design concept. Under this concept, permeable materials were again placed at the base of the pit, and part way up the side walls; although for the upper portion of the tailings, the material is placed directly against the natural geologic materials.

In the Key Lake case, the natural geologic environment provided the required bypassing of the tailings deposit to achieve the desired environmental performance while the base drain again facilitated pumping during operations to achieve consolidation objectives (Figure 2). In both the initial design of the facilities, and through more than several decades of operation, it is clear that although in-pit tailings facilities provide common benefits with respect to long-term geotechnical stability, it is critical that their design approach take into consideration site-specific conditions (geologic, hydrologic, geochemical, environmental, and operational constraints), as well as continue to apply adaptive management principles throughout operation and into closure planning to optimize performance. For example, at Rabbit Lake, deposition evolved from subaerial deposition (the initial deposition approach) to one where the tailings are placed under water cover to prevent further development of ice lenses within the facility (problematic given the cold northern climate). At Key Lake, the additional volumes available in the open pit allowed for co-milling of low-grade ore, while the subaqueous deposition approach allowed for a large reduction in the dewatering volumes requiring treatment and release. As each facility approaches the end of its operational life and enters the closure period, further studies continue to refine the eventual long-term closure plans and evolve optimal approaches to reduce downstream loadings to as low as reasonably achievable. ■

PART II

Operations and Monitoring

Integrated Monitoring Center Provides Continuous Real-Time Dam Monitoring, Minas Gerais, Brazil

João Antônio Teixeira Soares
Companhia Brasileira de Metalurgia e Mineração (CBMM)

Marcos Antônio Lemos Júnior
Companhia Brasileira de Metalurgia e Mineração (CBMM)

The Companhia Brasileira de Metalurgia e Mineração (CBMM) located in Araxá, State of Minas Gerais, in southeast Brazil is a worldwide leader in the supply of niobium products, which can transform materials into intelligent solutions, thus building a more sustainable and efficient future.

CBMM has six dams: four for deposition of tailings, one for accumulation of fresh water, and one for sediment retention. The structures are managed by a specialized technical team, responsible for the areas of design, tconstruction, operation, monitoring, inspection, maintenance, and compliance with legal requirements.

Aiming at continuous improvement and enhancement in safety, in April 2019, CBMM deployed its Integrated Monitoring Center (IMC). The IMC operates 24 hours a day, 7 days a week (Figure 1). A specialized technical team monitors and analyzes in real time all the information generated by the instruments and other monitoring tools of the dams. This same team is responsible for notifying the senior technical staff of any eventual emergency situation and for triggering the sound alert system (sirens) installed in the first 10 km downstream of the dams.

The technologies implemented to monitor the behavior of the dams are:

- A video monitoring system, consisting of high-resolution cameras and storage of images for at least 90 days
- Automatic monitoring of flow conditions in the area of the dam, made with water-level indicators, piezometers, and a V-notch weir
- Automated instruments to monitor the reservoir water level
- An automated meteorological station
- Automated monitoring of displacement using a Leica Robotic Total Station and interferometric synthetic aperture radar (known as *InSAR*) technology
- Microseismic monitoring with high-precision automated seismographs mesh in the entire CBMM area
- A supervisory system to monitor sensor status

Courtesy of CBMM

FIGURE 1 CBMM's Integrated Monitoring Center

CBMM has a specialized technical team to perform preventive maintenance and calibrate every instrument to ensure operational condition for all the monitoring technology.

Taking into consideration the particularities and monitoring needs of dams, CBMM developed its own platform to manage the monitoring results, customized to meet the specificity of each structure. Therefore, all information generated in the IMC is forwarded to this platform: from the location map of dams and graphical visualization of instruments to the more complex applications of machine learning, alert notifications, predictive statistics models, and the deviation management tools of the monitoring system. The applications and tools are called *operational controls*. The operational controls aim to prioritize and improve the technical analysis of the IMC team as discussed next.

PREDICTIVE STATISTICS MODEL AND DEVIATION NOTICE

The water-level indicators and piezometers of the dam have a predictive statistic tool, which:

- Interprets the generated information compared to the provided control levels
- Eliminates eventual outliers
- Generates a notification to the technical team if a deviation is found in the instrument readings, based on an established trigger by the senior technical staff, as seen in Figure 2

This analysis and elimination of eventual outliers is performed by the box-plot statistics model, a tool to evaluate the distribution of readings registered by the sensors, taking into consideration the tendency of the behavior of instruments, their history, and recurrence of readings.

The trigger notices are forwarded via email to the technical team, who immediately respond to a standard flowchart that adopts the dynamic of "question, response, action" to prioritize and optimize the technical evaluation of the IMC team. The flowchart answers are automatically sent to the senior technical staff.

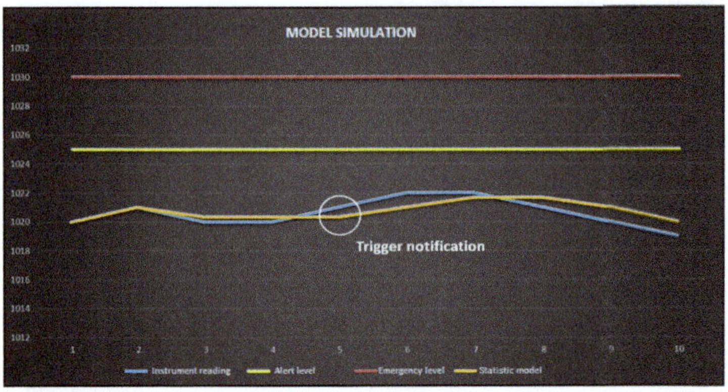

Courtesy of CBMM

FIGURE 2 Model simulation

Trigger notices are stored and are traceable with respect to date/hour of deviation, professional person responsible for filling in the flowchart, dam, instrument, control level reached, and occurrence status.

DAILY STATISTICS MODEL

In addition to automated monitoring, CBMM performs an operational routine of manual monitoring. This practice leads to higher sensitivity and knowledge for the technical team of the dams about the behavior of the structures.

Considering the automated readings of the water-level indicators and piezometers, statistical analyses of data are conducted to determine a daily representative reading for each instrument. These analyses are performed by the same statistics model described earlier, including the elimination of eventual outliers.

The purpose of this tool is to provide and compare both ways of gathering monitoring data that help the sensor calibration plan.

TRIGGERING EMERGENCY SIRENS

For the remote triggering of the sirens, IMC has two redundant systems, and a technical team duly trained to perform it. IMC also has an automatic triggering of sirens, based on parameters of displacement. To meet this demand, the automatic monitoring of displacement, performed by Robotic Total Stations, is integrated with the emergency sirens.

The integration of these two systems was carried out to comply with the triggering plan developed with the design company, where four control levels for vertical and horizontal displacements were defined: normal, attention, alert, and emergency. If the Robotic Total Station identifies an emergency level in all the superficial monitoring marks of the same transversal or longitudinal section of the dam, the command triggers the emergency sirens of the dam.

SOIL CONSERVATION SERVICE CURVE NUMBER CALIBRATION

The Soil Conservation Service Curve Number (SCS-CN) is a method for predicting the runoff volume behavior regarding a rainfall event. The curve number (CN) parameter is based on land management conditions and geology. However, Brazilian watersheds have a different geology and land management than the one used in the development of the method. Therefore,

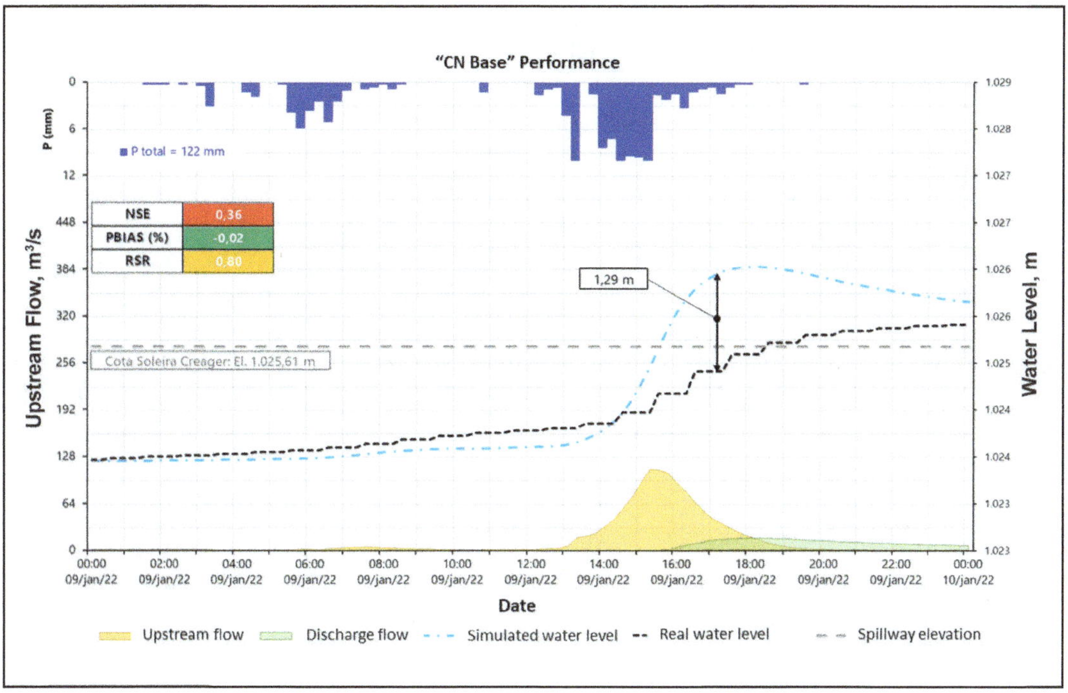

Courtesy of CBMM

FIGURE 3 Curve number base performance

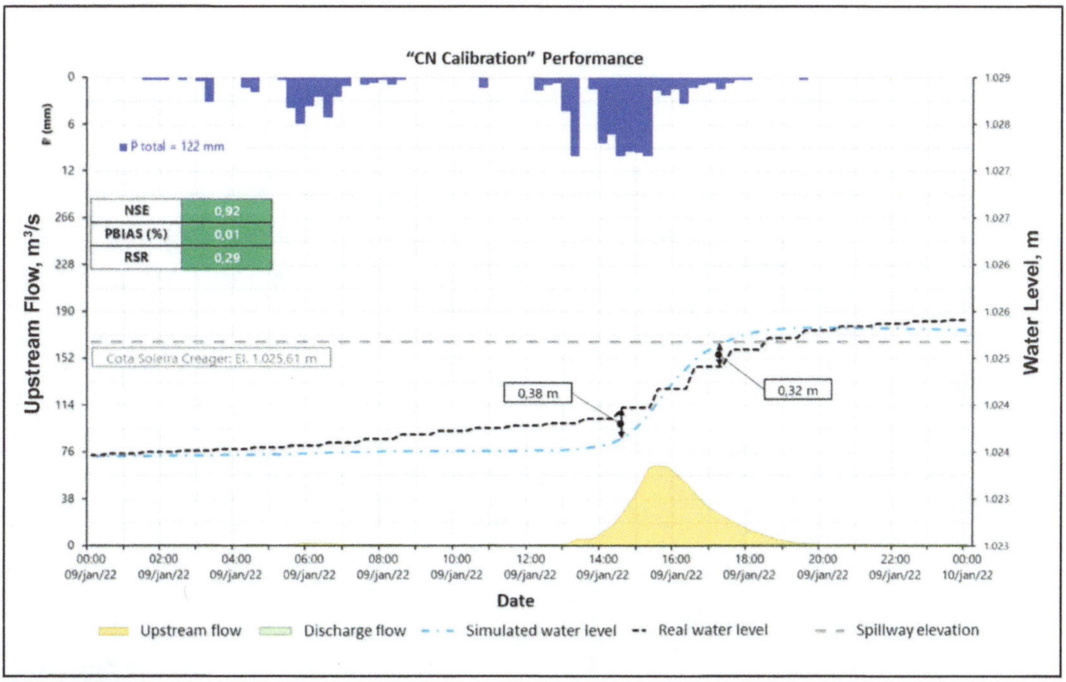

Courtesy of CBMM

FIGURE 4 Curve number calibration performance

CBMM decided to calibrate the CN using IMC's automated monitoring data such as reservoir water level, precipitation depth, spillway discharge flow, topography, and operational information.

The methodology used in the calibration was to determine the CN from the classic SCS-CN method, called *CN base* (Figure 3), and then evaluate its performance on the modeling, comparing the real water reservoir level (from the monitoring data) and the simulated reservoir water level given by the model. After that, the CN parameters that reach a better performance according to the most important events that occurred in the year, which were selected to be calibrated, are compared. The CN that reaches high adherence is called *CN calibration* (Figure 4).

Regarding the results (Table 1), it was possible to observe a reduction of approximately 20%, on average, on CN values. It is an important result that improved our hydrological safety analysis and the resilience of spillway performance, based on representative watershed physical parameters, adherent to its real behavior.

TABLE 1 Performance parameters results

Performance Parameters*	Base	Calibration
Precipitation depth, mm	122	
Started water level, m	1,024.15	
Maximum upstream flow, m^3/s	111.21	65.48
Maximum discharge flow, m^3/s	17.88	1.69
NSE (statistical parameters)	0.36	0.92
PBIAS (statistical parameters)	−0.02	0.01
RMSE (statistical parameters)	0.8	0.29
CN	73.2	59.5

*CN = curve number; NSE = Nash-Sutcliffe efficiency; PBIAS = percent bias; RMSE = root-mean-square error.

Courtesy of CBMM

SUMMARY

Every technology, application, and operational control that comprises the IMC are considered operation tools that support the engineering judgment of the dam management technical team. Such information does not override geotechnical knowledge and does not replace the operational routine of field inspections and monitoring.

New technologies for dam monitoring are frequently evaluated, and the IMC is constantly improving. Its operation undergoes standardization cycles based on routine management methodologies and periodic evaluations of its technical-operational level. It currently operates at maturity levels that meet and contribute to the best practices in monitoring dams. ■

Process Safety Approach to Tailings Management: Critical Control Reporting

Kimberly Finke Morrison
Newmont Corporation

George Afriyie
Newmont Corporation

Post-evaluations of notable tailings storage facility (TSF) failures reveal short-sightedness in recognizing and addressing certain defects related to site characterization, but also miss key attributes in operational practices, performance monitoring, managing change, and a lack of sustained review by subject matter experts. Newmont's corporate tailings management team within technical services developed a *process safety* approach to manage tailings management risks. Process safety may be defined as "the protection of people and property from episodic and catastrophic incidents that may result from unplanned or unexpected deviations in process conditions" (Center for Chemical Process Safety 2011) or "a blend of engineering and management skills focused on preventing catastrophic accidents" (Macza 2008).

To understand process safety, a comparison to occupational safety, often termed *personal safety*, is required. This distinction is illustrated in Figure 1. Personal safety typically provides good visibility into numerous incidents at the bottom of the pyramid to drive system improvements; while with process safety, the bottom of the pyramid is less visible and more complex to measure. Serious incidents, such as catastrophic failure of TSFs, have driven evolution of the process safety approach.

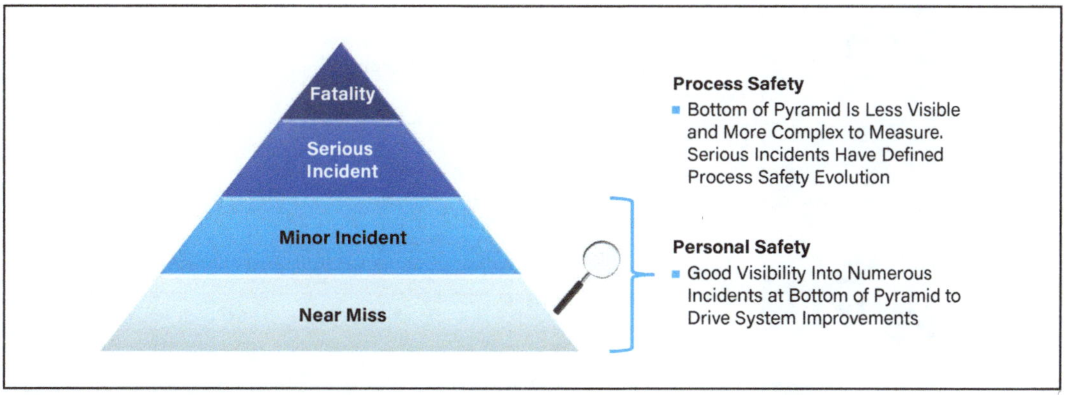

Adapted from Reimer 2021

FIGURE 1 Process safety versus personal safety

Within this overarching approach, a governance framework that utilizes a line-of-defense model has been implemented, which comprises four levels as shown in Figure 2 and provides clear definitions of the key roles involved in tailings management. To achieve the desired results from the implemented tailings governance framework, practical processes have been defined to place the right emphasis on performance expectations, operational controls, change management, and review processes. Site personnel from the first line of defense serve as major stakeholders in the implementation of the process safety approach. Corporate standards, guidelines, and procedures guide and support implementation of the framework.

The critical control report (CCR) is one of the key elements for the effective implementation of the company's governance framework. Critical controls are metrics that define the means to measure performance against various project objectives—primarily safety and risk management—and demonstrate the effectiveness of facility operation and performance. Critical controls are those that significantly influence the likelihood and/or consequence of an unwanted event. Those controls considered critical for TSF management are tracked at the sites and reported on a monthly basis by the site's responsible tailings facility engineer (RTFE) with support from the respective engineer of record (EoR). A notification and escalation matrix procedure has been developed to provide clear instructions for open lines of communication between the various levels of the organization to ensure timely notification of

First Line of Defense	**Site-based implementation** The site teams that actively manage the facilities include the site's general manager (GM); a responsible tailings facility engineer (RTFE), who is responsible for the integrity of the site's tailings facility (or facilities) and oversees their design, construction,and monitoring; and a dedicated engineer of record (EoR) and EoR team, who provide technical expertise and assurance that the tailings facility is safely designed, constructed, and operated. The regional leadership team (RLT) is responsible for the respective site team.
Second Line of Defense	**Systematic third-party review** This process includes independent tailings review boards (ITRBs) for all active operations and legacy sites with an "extreme" or "very high" consequence classification, senior independent technical reviews for other tailings facilities, and periodic dam safety reviews (DSRs) by an independent consulting team.
Third Line of Defense	**Functional accountability** An accountable executive (AE), who reports directly to the CEO, communicates regularly with the board and is accountable for the safety of Newmont's tailings facilities. The corporate tailings and dams team (CTDT), serves as a delegate to many of the AE's responsibilities, providing centralized leadership, oversight,and support for all tailings facilities (including inactive and legacy operations).
Fourth Line of Defense	**Leadership commitment** Newmont's executive leadership team (ELT) and the board of directors commit to the safe management of Newmont's tailings facilities. The board's safety and sustainability committee provides oversight, and the AE communicates with the board through the committee.

Source: Afriyie and Morrison 2023

FIGURE 2 Company's line-of-defense model for tailings governance

trigger exceedances or significant events to site, regional, and corporate leadership. The escalation goes through the ranks of the organization, up to the accountable executive (AE).

The CCR serves as a tool used to capture deviations from normal operations that may occur. Previously, the company deployed a CCR process that employed three reporting levels (i.e., stoplight approach with green, yellow, and red status indicators) and four categories for the critical controls, as documented by Morrison (2022); however, lessons learned through the process resulted in a change to the reporting framework (Afriyie and Morrison 2023). Now, the company employs four reporting levels and six categories for the critical controls. The reporting levels include:

- Acceptable situation (green status): Performance and operating requirements are within acceptable limits.
- Minor concern (blue status): Trends or changes observed that warrant review. Investigate and evaluate changes.
- Moderate concern (yellow status): Trends or changes persist that warrant action beyond current actions or practices being taken. Develop and implement action plan to mitigate further escalation.
- Major concern or threat (red status): Trends or changes that are determined to be a potential threat to TSF safety. Immediate action is needed.

A summary of critical control categories and the description for the reporting levels are shown in Table 1. Within the revised protocol, several key performance metrics that are not critical controls but are considered essential to regular TSF status reporting to the AE and other senior leadership team members have been included (e.g., slurry density). For more information on each of the critical control categories, see Afriyie and Morrison (2023).

In 2018, an Excel-spreadsheet-based system was initially used for preparation of CCRs, with each site preparing a single CCR to report on the overall "health" of tailings management at the site. The system was greatly enhanced in 2020 via launch of the company's internal TSF CCR guideline coupled with an online tool called *Sentinel*, which provides a robust database to manage information, support scheduling capabilities to increase accountability, and improve management of critical control performance. The dashboard is automatically sent to leadership, including the AE for tailings, providing visibility of tailings performance to the highest levels of the organization. Independent of the CCR process, unusual events or trigger exceedances relating to TSF performance are escalated via the company's tailings management escalation and communication matrix procedure, which ensures that the appropriate individuals are being notified and within the appropriate time frames. An example extracted from the procedure is illustrated in Figure 3 as it pertains to critical control No. 1 (i.e., exceedances of trigger levels).

As mentioned previously, the company first employed a four-level CCR, as discussed in Morrison (2022). The previous CCR process focused on four critical controls: (1) monitoring instrumentation (e.g., piezometers, inclinometers, deformation) against established trigger or threshold levels; (2) monitoring reclaim pond against operational criteria (e.g., location, volume, freeboard, beach width); (3) independent review and tracking of actions; and (4) change management related to design, construction, and operation. Additionally, the former process employed a stoplight approach for each of the controls (i.e., green, yellow, and red status

TABLE 1 Summary of critical control (CC) categories and the reporting levels

CC Category	CC Description			
	Green—Acceptable	Blue—Minor Concern	Yellow—Moderate Concern	Red—Major Concern/Threat
CC1—Performance monitoring	All trigger action response plan (TARP) measurements are below established threshold or trigger levels.	One or more TARP elements have exceeded established Threshold 1 triggers.	One or more TARP elements have exceeded established Threshold 2 triggers.	One or more TARP elements have triggered an emergency response level (Level 1, 2, or 3).
CC2—Operational management	All measured operational indicators or controls meet established criteria.	Recoverable delays or short-term deviation from established criteria.	Irrecoverable delays or deviations from established criteria leading to operational upset.	Irrecoverable delays or deviations from established criteria leading to temporary shutdown in operations.
CC3—Reviews and inspections	Inspections or reviews completed as per required frequency with Priority 3 and Priority 4 recommendations on schedule and/or complete.	Inspections or reviews not completed as per required frequency but scheduled. OR One or more Priority 3 actions overdue or not scheduled.	Inspections or reviews not completed as per required frequency and not yet scheduled. OR One or more open Priority 2 recommendations.	One or more open Priority 1 recommendations from reviews and inspections.
CC4—Change management	No design, construction, or operational changes.	Any design, construction, or operational changes have been reviewed and approved by corporate and other applicable personnel.	Planned design, construction, or operational changes are under review and have not been approved by corporate and other applicable personnel.	Design, construction, or operational changes have been implemented without review or approval from corporate and other applicable personnel.
CC5—Regulatory compliance	No regulatory actions, violations, or noncompliances associated with the TSF.	Regulatory action, violation, or noncompliance associated with the TSF. AND Action plan in progress and approved by the regulatory agency.	Regulatory action, violation, or noncompliance associated with the TSF. AND Action plan pending development and/or approval.	Cease-and-desist order received from regulator. OR Violation at the TSF with potential for a cease-and-desist order from regulator.
CC6—Emergency response plan (ERP) activation	No ERP activation for the TSF.	Not applicable (N/A)	N/A	ERP activated for the TSF (Level 1, 2, or 3 emergency).

Source: Afriyie and Morrison 2023

indicators). Over the years that this approach was followed, lessons were obtained that necessitated a change to the approach, as follows:

- The former CCR process was dependent on the personal biases of the personnel completing the CCR. The success of the system requires focused training with personnel involved to ensure that requirements are understood and consistently deployed.
- Over the earlier years of the CCR process, concerns such as activation of the emergency response plan and regulatory noncompliances were raised and were lumped into the change management bucket (i.e., critical control No. 4). Each of these elements has now been elevated to its own critical control to provide improved clarity on facility performance (i.e., via addition of critical controls Nos. 5 and 6).

	Trigger or Event	RTFE/P to GM Timeframe	Site to GM Mode	Site to EOR Timeframe	Site to EOR Mode	Site to CTDT Timeframe	Site to CTDT Mode	Site to RSVP/VPP Timeframe	Site to RSVP/VPP Mode	Site to AE Timeframe	Site to AE Mode	AE to COO Timeframe	AE to COO Mode
Performance Monitoring	Normal Conditions	N/A	CCR	N/A	N/A	N/A	CCR	N/A	CCR	N/A	CCR	N/A	CCR
	Threshold 1 - Investigate	N/A	CCR	<12 hours	Email	N/A	CCR	N/A	CCR	N/A	CCR	N/A	CCR
	Threshold 2 - Notify	<12 hours	Email	<8 hours	Email	<24 hours	Email	N/A	CCR	N/A	CCR	N/A	CCR
	Level 1 - Unusual condition, slowly developing	<2 hours	Phone + Email	<2 hours	Phone + Email	<8 hours	Email	<8 hours	Email	<24 hours	Email	N/A	CCR
	Level 2 - Potential failure situation, rapidly developing	Immediately	Phone + Email	Immediately	Phone + Email	<2 hours	Text + Email	<2 hours	Text + Email	<8 hours	Text + Email	<2 hours	Text + Email
	Level 3 - Urgent, failure is imminent or in progress	Immediately	Phone + Email	Immediately	Phone + Email	<2 hours	Text + Email	<2 hours	Text + Email	<2 hours	Text + Email	Immediately	Text + Email

Source: Afriyie and Morrison 2023

FIGURE 3 Excerpt from the escalation and communication matrix procedure

- Early identification of changing conditions is key to implementing the necessary corrective actions. To this end, an additional reporting level (blue—minor concern) was added to identify minor changing conditions that may not previously have been communicated or indicated, or were elevated to a higher concern (i.e., yellow status).
- Direct linkages to the site-specific trigger action response plans (TARPs) ensures that reporting on performance monitoring is consistent, replacing the previous critical control Nos. 1 and 2 (Figure 4).
- The CCR serves as a valuable source of information for executive leadership, including the AE. However, the former reporting process had a tendency to elevate risk concerns to red level when not necessitated, and resulted in numerous yellow level indicators, which can potentially lead to complacency around tailings management.

An approach for identification of deviations from normal conditions for the management of TSFs is presented. The CCR process allows for changes to be communicated to stakeholders inclusive of the AE on a monthly basis. Alignment with this critical control strategy across the operations and cross-functional disciplines is essential in ensuring consistent and accurate reporting of associated risk levels for the various TSFs within the company's global portfolio. In addition to the monthly CCR, an escalation matrix procedure for tailings management is used to report deviations in TSF performance that occur in a timelier manner. The escalation matrix ensures proactive reporting of changes to stakeholders at appropriate levels of the organization for effective mitigation, as required. The approach discussed herein has proven effective in placing the appropriate emphasis on performance expectations, operational controls, and change management practices in the management of risks associated with the company's TSFs.

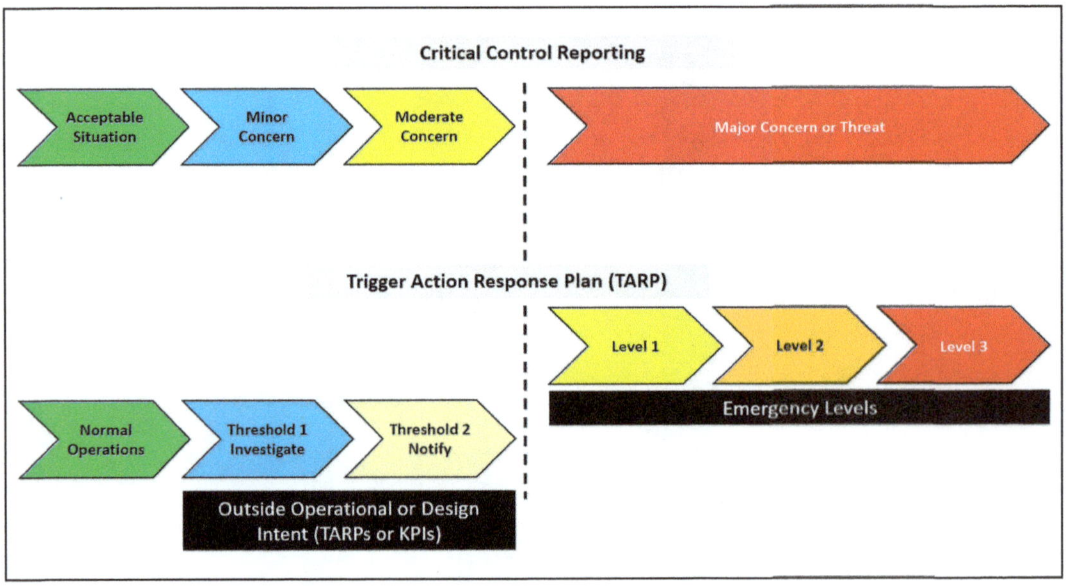

Source: Afriyie and Morrison 2023

FIGURE 4 Linkages between TARPs and critical control reporting levels, including key performance indicators (KPIs)

REFERENCES

Afriyie, G., and Morrison, K.F. 2023. Process safety approach for reviewing critical controls on tailings storage facilities. In *Proceedings of Tailings and Mine Waste 2023*. Edited by J. Goodwill. Vancouver, Canada: Norman B. Keevil Institute of Mining Engineering, University of British Columbia. pp. 1055–1067.

Center for Chemical Process Safety. 2011. *Guidelines for Auditing Process Safety Management Systems*, 2nd ed. New York: American Institute of Chemical Engineers.

Macza, M. 2008. A Canadian perspective of the history of process safety management legislation. Presented at the 8th Internationale Symposium, Programmable Electronic System in Safety-Related Applications, Cologne, Germany, September 2–3. www.acm.ca/sites/default/files/news-article-pdfs/A%20Canadian%20Perspective%20of%20the%20History%20of%20PSM%20Legislation.pdf. Accessed March 2021.

Morrison, K.F. 2022. Process safety approach to tailings management: critical control reporting. In *Tailings Management Handbook: A Life-Cycle Approach*. Edited by K.F. Morrison. Englewood, CO: SME.

Reimer, A. 2021. *Process Safety vs. Personal Safety* [slide presentation]. https://pdf4pro.com/view/process-safety-vs-personal-safety-home-enform-473aed.html. Accessed March 2021.

Sand Tailings Deposition Through L Lines on Slopes on High Embankments, Cerro Verde Mine, Arequipa, Peru

Henry Antonio Arapa Alvarez
Freeport-McMoRan

Nelson Roldan Aranibar
Freeport-McMoRan

Juan José Navarrete Cabrera
Freeport-McMoRan

Juan Carlos Tacca Apaza
Freeport-McMoRan

Diego Bustinza Martinez
Freeport-McMoRan

During 2006 and 2015, Sociedad Minera Cerro Verde (SMCV), a subsidiary of Freeport-McMoRan, commissioned and started the operation of the Quebrada Enlozada and Quebrada Linga tailings storage facilities (TSFs). A zoned rock-fill starter dam was used to initiate operation for each of the TSFs. The TSFs' embankments are continually raised with compacted cycloned tailings underflow using the centerline method.

The tailings coarse sands are transported and discharged from the classification plant (cyclone tailings underflow) to the embankment crest by a main distribution line and a T-shaped spray bar system mounted on supporting and lifting pipe structures called the *jacking header system*. The initial design included intermediate lines on the slopes with a T-shaped spray bar system, but these intermediate lines interfered with the operation, construction, and raising of the embankment because they did not use a jacking header system to accommodate the high rates of raise. The originally designed intermediate lines also posed greater safety risks because of the continuous handling of lines and longer times caused by the change of zones. The sand that is discharged by the spray bar system gravity flows from the crest down the slope 150 m and then is pushed by a dozer to approximately 250 m. However, this is not an adequate distance to cover the entire slope.

A system was developed that extended sand placement on the slope to up to 1 km. SMCV implemented a system of L lines that consist of branching high-density polyethylene (HDPE) lines from the coarse pipeline of the jacking header by extending them along the slopes to lengths of 250, 500, and 750 m and then curving them at approximately 150–200 m to form an L shape with spigots spaced every 2.2 m (Figure 1).

The pressure ratings of pipes and fittings were considered to control pressures at the discharge points. Operational procedures were established to eliminate the use of constraints or open valves in the branch of the L lines and spigots before they receive sands. This is to prevent a high-pressure static flow state in the lower points of the L and to avoid sanding or plugging of lines by sedimentation. With this operation, the vertical part of the L initially flows partially

Courtesy of Sociedad Minera Cerro Verde

FIGURE 1 Sand tailings deposition through L lines at slope

TABLE 1 L system advantages and disadvantage

Advantages	Disadvantage
Reduced need of heavy machinery: With this system, discharge points can be positioned in the area or at any point of the slope where more sand is required, thus minimizing the work of heavy machinery by not having to push the sand to its final location. **Less fines generation:** When discharging sand by gravity from the crest to its final location, the degradation of fines is minimal as opposed to discharging sand at the head and pushing it with heavy machinery to its final location, where the sand continues to disintegrate and generate greater fines content. **Homogeneous discharge:** It is easier to assemble lines with diameters smaller than 15.24 cm in each specific position and at the same level of the slope, which allows a homogeneous discharge without generating erosion that affects the layers previously built.	**Continuous installation of L lines:** By using multiple discharge lines, it is necessary to assemble and relocate mobile lines throughout the operational life. It is also necessary to have the equipment for the preparation, installation, and maintenance of high-density polyethylene (HDPE) lines.

full and is later pressurized but does not exceed the pressure rating of the HDPE. The HDPE lines must be changed every 9 to 10 months because of the high wear. The operation of the L system provides three advantages and one disadvantage shown in Table 1.

In summary, for transporting sands in slopes of high embankments, conventional HDPE lines in L shapes can be used on slopes with inclinations of 28%, operating at design flows, with smaller diameters to increase friction loss and obtain operating pressures within the pressures of design. This assures free sand discharge without constraints in the L line to prevent the static hydraulic load and to significantly reduce the risk of pipe breakage and slope erosion. ■

Management of Wet Cycles at Mine Facilities, Saskatchewan, Canada

Brian Bellmore
Barr Engineering Company

Art Kalmes
Barr Engineering Company

The management of mine water during extended wet cycles can present more challenges than relatively short-term regulatory design storms. This case study shows how this fact was the driver for storage pond management at facilities where discharge is limited.

A Canadian mining company operates a group of underground mines in Saskatchewan. Each mine manages water in storage ponds that capture stormwater runoff from the tailings storage facility (TSF). The water volume fluctuates over time with changes from precipitation, evaporation, and reuse for mill ore processing. The excess water is disposed by injection into a deep, confined aquifer because it cannot be discharged to surface waters. Provincial regulations require the mines to design containment dams to store a regulatory storm that is based on Canadian Dam Association guidelines for the high hazard dam consequence classification. The regulatory storm is one-third between the 0.1% annual exceedance probability (AEP) and the probable maximum flood and has been interpreted to have a 24-hour storm duration. The 24-hour regulatory storm depth is 297 mm.

The mine designs the storage pond size and manages injection-well trigger levels to maintain the regulatory storm storage. However, the site disposal is limited to the injection-well capacity, which is relatively small compared to the potential TSF stormwater runoff as it would take nearly 200 days to draw down the regulatory storm by injection. The mines were concerned that extended wet cycles can raise pond levels into the regulatory storage zone for long periods of time. Operating the ponds at high levels reduces storage for the potential regulatory storm and increases risk of dam-wave erosion and stability issues.

The mine owner commissioned a water-balance assessment to review the ability of each site to dispose of the 1% AEP (100-year return frequency) and larger events below the regulatory storage zone prior to April 1 the following year. This would limit pond operation at higher levels and maintain available storage for the higher precipitation months during spring and summer. The study looked at historic rainfall records for the region to determine precipitation frequency distributions for wet cycles of up to 1 year. Figure 1 shows how the rainfall depth varies with wet cycle duration. This figure illustrates that despite being far more frequent than the regulatory storm, 1% AEP events that exceed 30-day durations produce more rainfall than the regulatory storm. The mines need to have adequate injection capacity to prevent

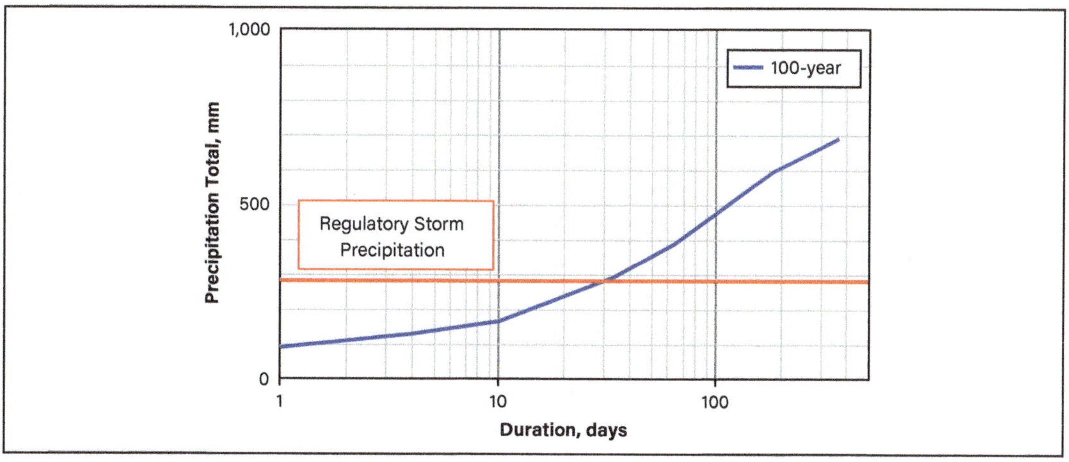

Courtesy of Barr Engineering

FIGURE 1 Precipitation depths versus wet cycle duration

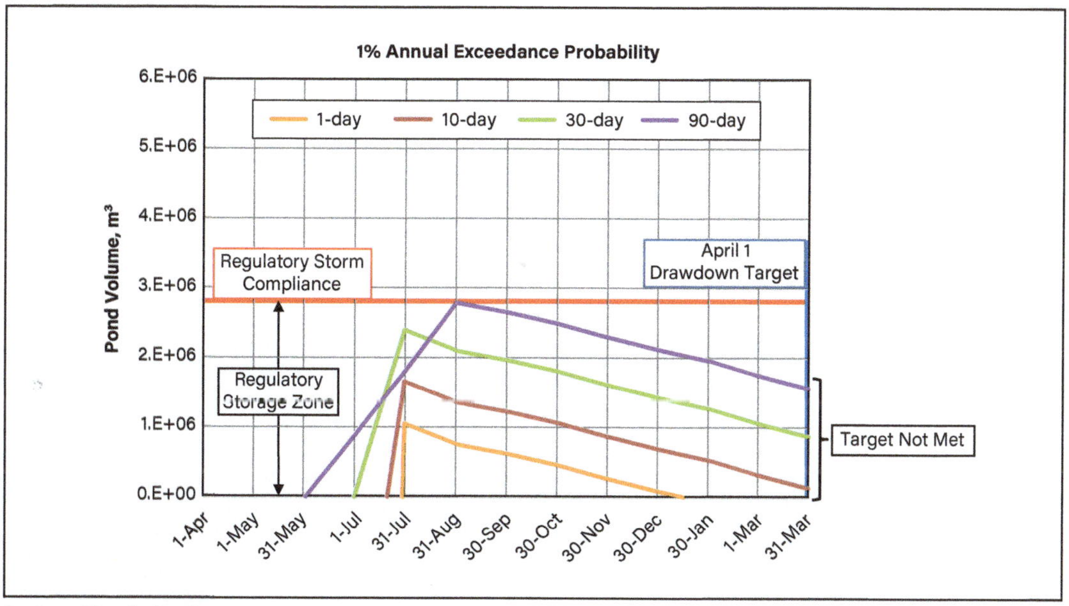

Courtesy of Barr Engineering

FIGURE 2 Pond volume change during various wet cycles

overtopping and to reduce pond levels after these more frequent long-duration wet cycles. A similar analysis may be done for more extreme events.

A water-balance model was created for each mine with site data and the precipitation frequency distributions. The model assumed average monthly precipitation after the storm occurred and full injection throughout. Figure 2 shows the modeled stored pond volume from one mine site where multiple 1% AEP wet cycles were evaluated. This assessment demonstrated that the injection wells could not dispose of the surplus water prior to April 1 the following year for wet cycles in excess of approximately 10 days. The stored pond volume approached regulatory compliance even though the modeled events were more frequent than the regulatory storm.

The assessment identified the potential risk prior to it being a situation that could threaten the dam. The assessment results were used to adjust operating levels so more storage for wet cycles is maintained. Mine management also used the results to support the need for increased injection capacity at several mines. The resulting mitigations have reduced the identified risk and helped continue the safe operation of the facilities.

Mine facilities that cannot freely discharge excess water face the challenge of maintaining storage for potential large storm events and prolonged wet cycles. Regulations typically dictate that mines maintain pond storage for a specific event. The mine owner assessed other situations that could expose the mine sites to risk by performing non-mandated assessments to meet the internal guidelines for operations. In this case study, the drawdown time assessment indicated that although the mines were meeting regulatory storage requirements, more frequent wet cycles posed a risk above the owner's tolerance that required mitigation. Limited discharge and zero discharge mines should assess their ability to store and dispose of runoff from long-duration wet cycles to evaluate their potential risk. ■

Modeling Overland Flow of Tailings Slurries

Ben Sheets
Barr Engineering Company

The overland flow of thickened tailings and other non-Newtonian fluids (e.g., fines-dominated tailings slurries, thickened or flocculated tailings, and various muds) is of great importance in many mining applications, from tailings flow into a basin to embankment breach analysis. However, comprehensive modeling and prediction of the flow patterns of such slurries is difficult because of the complexity of the equations describing their fluid dynamics. Fortunately, valuable insight and useful first-order predictions can be gained from a relatively simple modeling approach that captures the fundamental forces governing slurry flow.

Slurries of this nature are well described by classical "lubrication theory" (Liu and Mei 1990), in which the momentum equations can be significantly simplified. The Barr Engineering team developed and applied a simple model based on lubrication theory for non-Newtonian slurries that delivers rapid determination of flow characteristics (e.g., thickness, slope, and velocity) as a function of easily measured information: slurry material properties (density, yield stress), boundary conditions (tailings discharge rate, subaerial/subaqueous), and basin geometry (basement slope, locations of feed points). The model provides insight into the transient behavior of slurry flows interacting with their surroundings, and thereby allows predictions of slurry behavior over a range of scales, from laboratory to field-pilot to full-scale tailings deposition. The model is ideally suited for predictions in environments where relatively few field-scale data are available for model calibration and validation, and provides critical insight for guiding tailings basin design, developing deposition strategies, planning long-term basin management, and even predicting tailings runout from embankment breaches.

The model was validated against a Canadian oil sands flocculated tailings pilot test. Surface elevation LiDAR profiles of the tailings flow were obtained during a constant-discharge slurry deposition event from a single spill box on the edge of a 40-m × 40-m cell with a uniform basin floor slope of 1.7%. The team modeled (*hindcast*) the depositional event using a two-dimensional version of the model, applying the observed slurry yield stress and density. Figure 1 shows a comparison between model predictions (open circles) and observed slurry surface profiles (solid lines) at 2 minutes, 5 minutes 50 seconds, and 11 minutes after the beginning of the test. Overall, the model nicely captures the first-order behavior of the slurry: The curvature of the profile near the slurry toe matches the observed, with modest underprediction of flow thickness near the spill box, where the two-dimensional model does not fully capture the turbulent dynamics and rapid lateral expansion of the flow.

The model was subsequently applied to predict commercial-scale operations to provide mine operators with physically reasonable bounds on slurry flow thickness and velocity that might be expected with the basin geometry and various slurry properties. These predictions were a valuable guide for tailings deposition planning and long-term tailings management

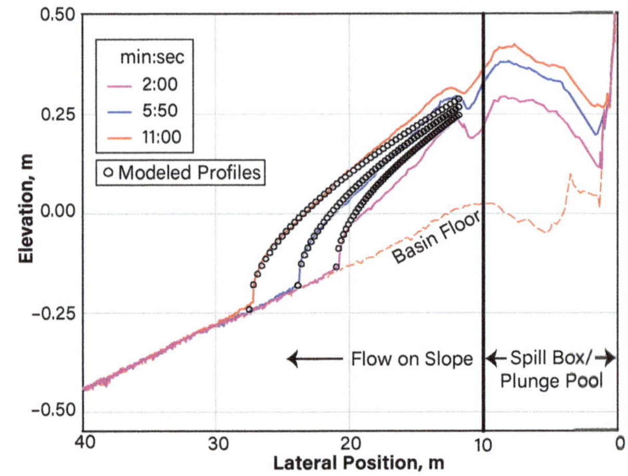

Adapted from Swenson et al. 2014

FIGURE 1 Comparison of measured pilot test thickened tailings flow profiles with model results

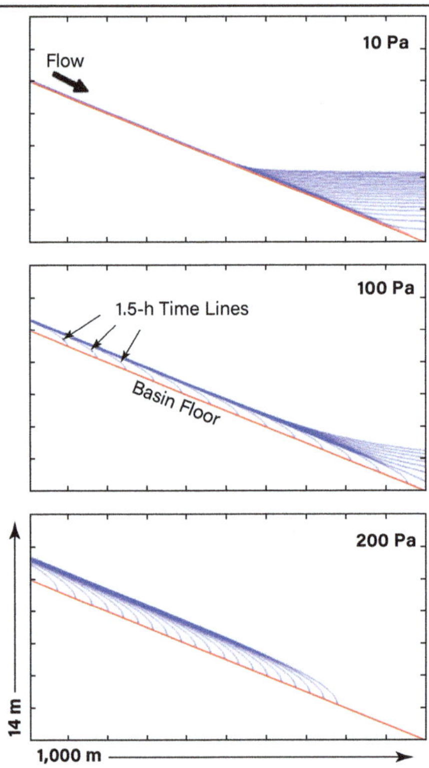

FIGURE 2 Plots of model results showing variable flow depth and runout distance as a function of thickened tailings yield stress

strategy. Figure 2 shows two-dimensional model predictions of deposition patterns associated with varying tailings slurry yield stress. Weak slurries (10 and 100 Pa) run out to the deepest portion of the impoundment, while thicker, stronger slurries (200 Pa) tend to deposit high on the basin slope.

This simplified modeling approach can save considerable resources by narrowing the range of feasible tailings mixtures and deposition strategies, thereby avoiding expensive field trials of tailings management options that would not prove viable. This type of model is also valuable for estimating velocity, distance, and thickness of slurry runout resulting from tailings embankment breaches. ■

REFERENCES

Liu, K.F., and Mei, C.C. 1990. Approximate equations for the slow spreading of a thin sheet of Bingham plastic fluid. *Physics of Fluids A: Fluid Dynamics* 2:30–36.

Swenson, J.B., Sheets, B., Kolstad, D., et al. 2014. Semi-analytical modeling of thickened tailings flows. Presented at the Fourth International Oil Sands Tailings Conference, Lake Louise, Alberta, Canada, December 7–10.

Modifying a Conceptual Site Model Through Additional Data Analysis and Visualization

Denise Levitan
Barr Engineering Company

Nathan Brandner
Barr Engineering Company

For many decades, a potash operation in Saskatchewan, Canada, has maintained storage impoundments for mine tailings and process-related mine water. The mine water is a concentrated aqueous solution, with most of the total dissolved solids (TDS) consisting of sodium and chloride. Although the facility has a system of slurry walls and collection/diversion ditches, the mines monitor the surface water and groundwater around the tailings storage facilities (TSFs) for releases.

Groundwater on the western side of the facility has been measured as having concentrations of TDS, magnesium, sodium, chloride, and sulfate above provincial drinking water quality objectives. In addition, electromagnetic surveys have shown high electrical conductivity in that area. For these reasons, studies were undertaken to investigate the extent of the high TDS concentrations in groundwater and whether the elevated concentrations were related to the TSF.

Chloride and TDS concentrations in groundwater monitoring wells were used to interpret the transport of high-TDS mine water moving along three interpreted zones of elevated TDS west from the facility (Figure 1). The zones of elevated TDS concentrations corresponded to the areas of high conductivity identified by the geophysical surveys. These results suggested that mine water could be emanating from the TSF and potentially require remediation. However, the interpreted orientation of the high-TDS groundwater and apparent flow paths were perpendicular to the direction of groundwater flow in the area. The fundamental conflicts in these interpretations led to a re-examination of the conceptual site model for the distribution of potential mine water releases.

Although the data examination and visualization made it clear that TDS concentration alone was not a reliable indicator of mine water release, there was still concern about elevated chloride concentrations. To further evaluate high TDS and chloride concentrations, several background area wells were installed northwest of the facility, beyond where mine water impacts could occur (Figure 2). The background wells had highly variable chemical compositions. One of the wells had an elevated TDS concentration with magnesium and sulfate as the

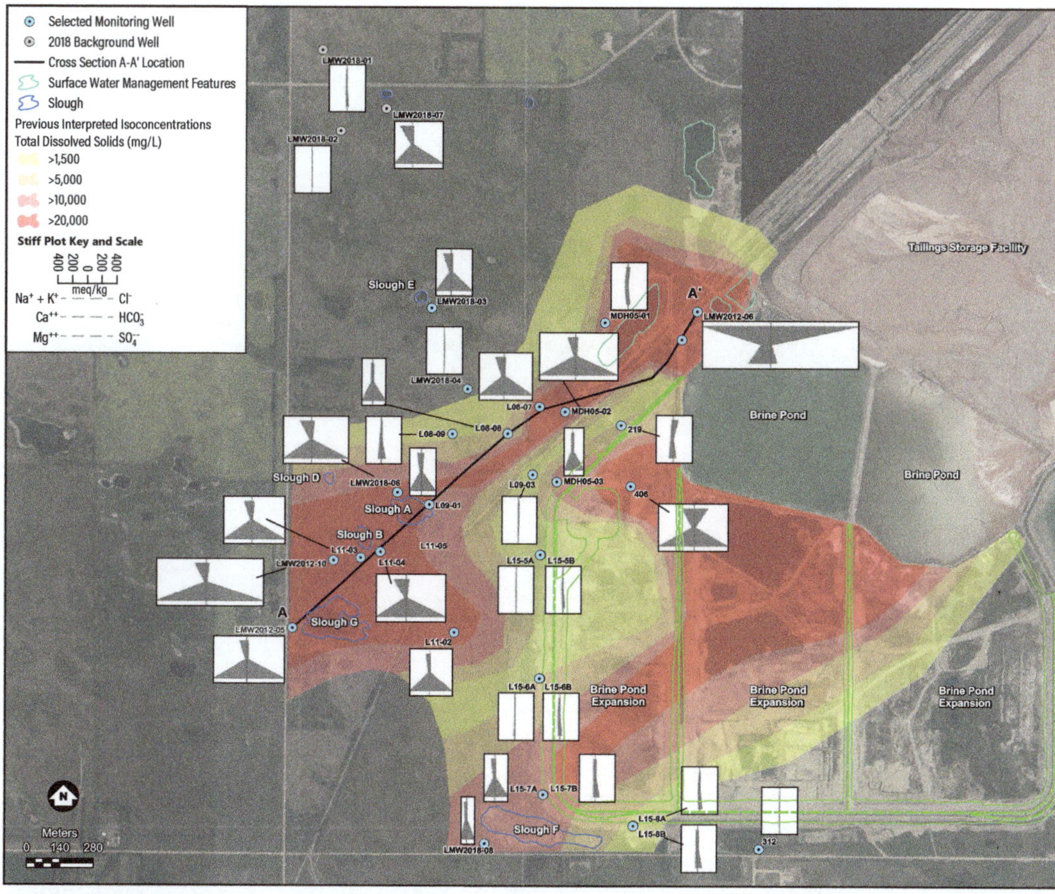

Courtesy of Barr Engineering

FIGURE 1 Plan view of the western side of the facility. The previously interpreted high total dissolved solids (TDS) flow paths are shown as an isoconcentration heat map. Stiff diagrams are shown for selected wells using a common scale. A black line indicates the location of the cross section shown in Figure 2.

dominant ions with chloride concentrations above drinking water objectives, similar to the high-TDS wells west of the facility.

The results of the background groundwater investigation (major ion chemistry) were supported by the scientific literature on prairie sloughs throughout the North American Great Plains. These sloughs are small, shallow, and sometimes seasonal wetlands found throughout the region. Seasonal cycling in precipitation/runoff and evapotranspiration result in accumulation and mobilization of salts, especially sulfate-bearing phases such as gypsum. The long-term dynamics of these features, which are common on the property west of the facility, likely control the TDS concentrations and compositions in shallow groundwater rather than mine water migration from the site.

The conclusions from this investigation led to a better understanding of the lateral distribution of potential TSF-related impacts versus background groundwater chemistry. Presentation of the data, including the visualizations of different water compositions, to the provincial regulatory agency representative allowed the facility to propose an optimized annual groundwater

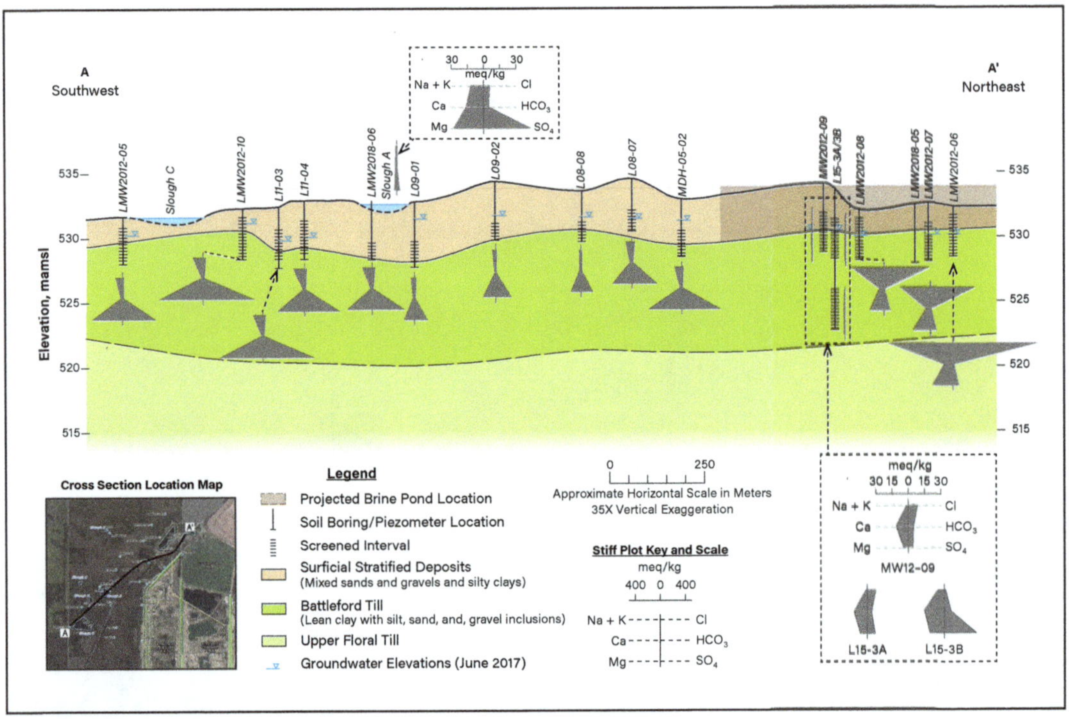

Courtesy of Barr Engineering

FIGURE 2 Cross section west of the tailings storage facility (TSF) (TSF on right side). Stiff diagrams show sodium- and chloride-dominated groundwater with high concentrations of total dissolved solids (TDS) near TSF at right and magnesium- and sulfate-dominated groundwater with high TDS concentrations at left.

monitoring plan with reduced sampling frequency and annual sample count. This study underscores the importance of critical evaluation and comparison of various data types, including background levels, from a particular site and the iterative approach to developing and supporting robust conceptual site models for optimizing project outcomes. ■

Precipitation Forecasting for Tailings Storage Facility Water Management

Homin Kim
Freeport-McMoRan

Brant Whiting
Freeport-McMoRan

Understanding precipitation accumulation patterns is critically important to ensure proper infrastructure design, water management planning, and risk management, and reliably inform decision-making for mine management. Statistical analysis of historical data is often used to forecast precipitation, but exclusive use of on-site weather station data is often insufficient because it does not consider regional climatology, and the period of record is usually limited. Further, while event-based risk (e.g., overtopping potential caused by a 100-year, 24-hour event) can be readily quantified, longer-term patterns are often only vaguely defined and generally not explicitly tied to probabilistic outcomes (e.g., What if there is a wet year? What if there are two wet years in a row because of El Niño?). Because of this, the ability to characterize both short- and long-term historic precipitation patterns and incorporate this into probabilistic forecasts using fewer statistical models and maximizing the information gained from existing data is of great interest. This case study generally explains how to overcome this constraint via the development of future daily precipitation sequences with exceedance probabilities for multiyear durations, and how these can be used to support decisions on tailings management. While there are several methodological advantages to this approach, ultimately, these improvements facilitate clear and reliable estimates of long-term, multiyear risks to decision-makers.

In this case study, site management needed to know not only what level of event-based precipitation accumulation would cause runoff volumes to encroach on established freeboard requirements in a tailings storage facility (i.e., event-based 100-year, 24-hour, 72-hour probable maximum precipitation [PMP]) but also how multiyear precipitation patterns (and their associated probabilities) could create adverse conditions that could impact operations and dam safety.

The facility evaluated in this case study has one on-site weather station, with a period of record limited to 12 years. Regionally, there are 17 weather stations near the area of interest that are candidates to be incorporated into model forecast scenarios. After review of the 17 regional weather stations, 2 regional weather stations near the operating mine were chosen, using criteria such as a long period of record (more than 100 years at each weather station), data quality, reasonable distance from area of interest, and similar climatology. Following the selection of these quality-controlled regional weather stations, daily precipitation levels from the selected stations were aggregated into non-overlapping *n*-year duration totals (e.g., 1-year, 2-year, up to

TABLE 1 Depth-duration-frequency example

Total Precipitation as a Function of Duration and Exceedance Probabilities (EPs)					
EP, percent	1 Year, mm	2 Year, mm	3 Year, mm	4 Year, mm	5 Year, mm
50	284	569	848	1,130	1,433
20	358	673	983	1,316	1,618
10	396	732	1,062	1,427	1,712
4	437	795	1,151	1,554	1,808
2	462	836	1,207	1,641	1,864
1	485	871	1,260	1,722	1,910
0.5	505	902	1,306	1,798	1,948

Courtesy of Freeport-McMoRan

5-year accumulation horizons). Goodness-of-fit and correlation measures between aggregated regional data and on-site measurements were used to identify the best distribution for developing the exceedance probability (EP) curves. These curves were used to derive a depth-duration-frequency table: Table 1 shows cumulative precipitation scaled to on-site conditions for various durations and EPs. For example, in a 5-year duration, the data (more than 200 years of station data) indicate only a 0.5% probability of exceeding 1,948 mm of total precipitation.

As a final step, 30 separate independent daily precipitation time series were extracted from the historical data for each EP and future duration based on how closely the cumulative precipitation in the sequence matched the respective EP precipitation total. Although the selection of 30 time series in this example was considered to be sufficient by the authors to capture the general variability of accumulating precipitation to a given total, more sequences could be developed, based on the number of station-years available, and in some instances the number of sequences may be limited by the available gauge data periods of record.

An illustration of the 5-year and 0.5% EP precipitation values is shown in Figure 1. The selection of 30 different patterns provides a robust and data-based method of forecasting to account for variable temporal patterns to accumulate 1,948 mm and serves as a strong foundation for risk management and decision-making.

The technical benefits of such an approach are that patterns present in the long-term record near a site of interest can be used, thereby making it no longer necessary to explicitly model the independence and time-series correlations of precipitation patterns (e.g., Markov chains, independent monthly precipitation probability distributions) that are commonly used in water balance modeling forecasts.

From a management perspective, the first benefit of this approach is that site management was able to evaluate scenarios associated with different forecasted periods and their associated EPs as shown in Table 1. The EP explicitly defines the probability of accumulating a certain amount of precipitation over a given time frame that is based on many station-years of data. A second benefit is that the EP and corresponding duration can be modeled using a variation of temporal patterns (i.e., 30 sequences) to assess the systematic potential of exceeding a given criteria, for example, freeboard. Finally, this methodology allows the regional data to be used to their utmost and removes a requirement to statistically model climatologic patterns in model inputs. For example, this method intrinsically accounts for month-to-month and year-to-year correlations because of extended dry/wet periods caused by macroscale climate events. These patterns are very difficult to explicitly model, and any month-to-month and year-to-year

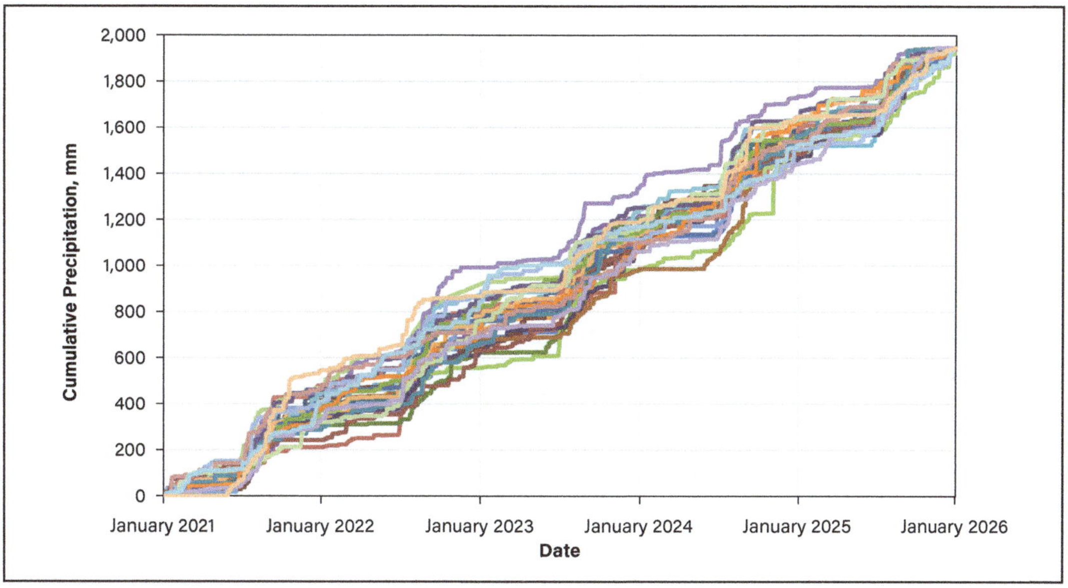

Courtesy of Freeport-McMoRan

FIGURE 1 Thirty precipitation patterns to accumulate 1,948 mm of total precipitation for 5 years and a 0.5% exceedance probability

TABLE 2 Summary of all 1,050 simulations with the number of precipitation patterns from 30 sequences by three categories

Category	Likelihood of Exceeding Freeboard Criteria				
	Unlikely		Potentially		Likely
Exceedance Probability, percent	1 Year	2 Years	3 Years	4 Years	5 Years
50	1	1	2	2	0
20	1	1	0	4	1
10	1	1	1	0	3
4	1	1	2	1	1
2	3	2	1	1	2
1	5	4	2	4	1
0.5	9	4	5	6	1

Courtesy of Freeport-McMoRan

patterns that emerge from modeling the climate series as independent monthly probabilities would only be random rather than part of a longer-term signal that is present in the regional data.

To address the concern of site management regarding the likelihood of exceeding a freeboard limit, seven recurrence probabilities and five time horizons (35 scenarios), shown in Table 1, were modeled, each using all 30 precipitation scenarios in GoldSim, version 12.1 (2020). This required 1,050 simulations. Furthermore, the simulations included a 72-hour PMP event during the later months of the simulation period.

The summary of simulations can be seen in Table 2. To determine freeboard exceedance, there are three categories, which are Unlikely in green, Potentially in yellow, and Likely in red. Unlikely is used when the average of 30 sequences does not exceed the criteria, and fewer than

5 sequences do exceed the criteria; Potentially is assigned when the average of 30 sequences does not exceed the criteria, but a total of 5 or more individual sequences exceed the criteria; and Likely represents the condition when the average of the 30 sequences exceeds criteria. Of the 35 scenarios, all but 4 were determined Unlikely to exceed the minimum freeboard criteria of 1.5 m. The remaining four scenarios were estimated to potentially exceed the criteria. For the four potential cases highlighted in yellow, the numbers given represent the number of sequences out of 30 that exceed the criteria: The overall average freeboard for the 30 sequences did not exceed the criteria in these cases.

Together, these results show that the current operating practices at the site are adequate to safely manage extreme multiyear precipitation accumulations combined with the 72-hour PMP for all durations at the 2% EP and greater. For the 1% and 0.5% EPs, the results show there is a potential for the freeboard criteria to be exceeded (e.g., instead of 1.5 m of freeboard, the facility only has 1.3 m of freeboard) for the 1-, 3-, and 4-year durations, but only after a very low probability (1%, 0.5%) precipitation pattern and a rare extreme storm event, that is, PMP.

Regional precipitation analysis is designed for use in generating daily precipitation patterns that are representative of future precipitation levels at a site. Utilizing data from regional weather stations overcomes the limitations of on-site data and quantifies the uncertainty over various durations. More importantly, the results in this case study indicate that the proposed approach reflects actual daily precipitation patterns and can be used to quantify risk going forward. Taken together, these findings significantly improve the capacity to forecast precipitation and can improve water management practices in the planning stages. This enables site managers to make proactive and informed decisions in consideration of precipitation accumulation patterns as well as extreme events, and their combinations. ■

REFERENCE

GoldSim. 2020. GoldSim 12.1 [software application]. Seattle, WA: Gold Technology Group.

Quantifying Uncertainty in Groundwater Heads Beneath a Tailings Storage Facility, Arizona, United States

Michael Gabora
Lake Resources

Eduardo DeSousa
DHI Water and Environment

Blair Douglas
BHP

BHP is working to improve groundwater flow model (model) forecasts utilized by tailings engineers to support tailings storage facility (TSF) engineering designs and operational decision-making. Often a single prediction, or at best a range of forecasts from a standard sensitivity analysis (e.g., ASTM D5611-94), is considered for these purposes. However, modern uncertainty analysis techniques provide an opportunity to re-evaluate what type of modeling outputs are used to support TSF management. The risk and consequence profile of many TSFs and adopting approaches to evaluate the likelihood of a specific occurrence of an unwanted event, such as an unacceptable pore pressure, provide decision-makers with the best information possible.

Ideally, this information is provided as a probabilistically derived ensemble of results constrained by the hydrogeologic understanding of the system. For example, the response of groundwater levels to a hydraulic stress can be evaluated by many (e.g., hundreds or thousands) constrained realizations of a model. Doing so permits the development of a cumulative distribution function of groundwater level results (Figure 1), which allows for stability analyses to be completed based on a probabilistically derived ensemble of results (e.g., 95th percentile). These types of results can be used to predict the likelihood of a particular pore pressure at a given location.

Rather than the more typical task of modeling seepage *out of* the TSF, in this case study we report on what was learned by modeling the impact of water flowing *into* the TSF domain, which can have implications for slope stability. The uncertainty analysis allows us to interrogate many model realizations that adequately match historical data at observation locations. The different versions of the calibrated models predict significantly different heads distal to the measured data, in areas where no observation data are available. The implication is that stability analyses can have varying outcomes depending on which realization of the calibrated model is used for the pore pressure analysis.

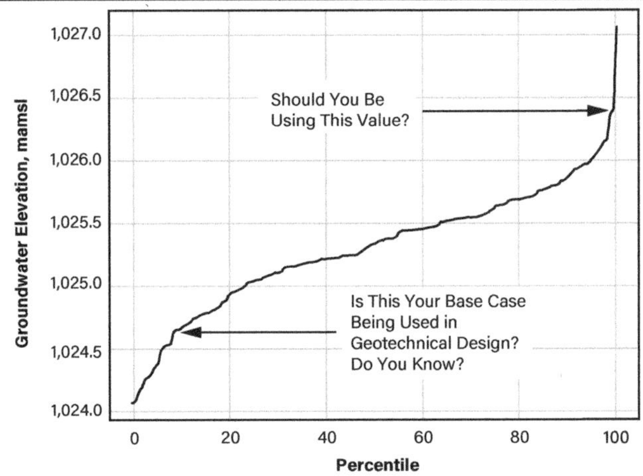

Courtesy of BHP

FIGURE 1 Example of a cumulative distribution function for groundwater elevation at a model observation point located at considerable distance from observation piezometers

The TSF selected for analysis is in a semi-arid region of Arizona and receives approximately 445 mm/y (millimeters per year) of precipitation, on average, and is subjected to significant upgradient catchment areas. During storm events, rainfall runoff infiltrates permeable channel alluvium upgradient of the TSF, resulting in seasonal ponding within the TSF basin, particularly along the alluvial channels. These alluvial drainage channels extend beneath both the impoundment and embankment and, during large storm events, hydraulic heads in the alluvium can increase up to 7.5 m in piezometers completed in the underlying alluvium, which is spatially and vertically limited.

The existing TSF groundwater model used as part of this study does not explicitly simulate the TSF, as it was originally developed to evaluate flow and transport of seepage from the facility. The existing model was modified from zone-based hydraulic properties to spatially variable hydraulic properties (Figure 2), known as a *continuous parameter field* (Doherty et al. 2010). A continuous parameter field is a prerequisite for exploring how heterogeneity in TSF foundation materials affects predicted groundwater heads.

History-matching the model to observations was completed for steady-state conditions and to transient responses over a 517-day period (January 2017 to May 2018), during which water levels rose 6.0 to 7.5 m beneath the TSF from a 34-mm precipitation event. PEST (Watermark Numerical Computing 2021) was used to facilitate the history-matching and adjust parameters within a range of values derived from extensive field characterization.

As part of the history-matching process, PEST generates sensitivity matrices (i.e., Jacobian matrices), which contain sensitivities of selected model outputs (e.g., field measurements of piezometric heads or fluxes) to parameter values. These sensitivities are used to calibrate the model through an iterative process in which parameter upgrades are tested, and model calibration is expected to progressively improve through several iterations until a minimum value for the objective function is obtained. During the uncertainty analysis, the parameters (or parameter combinations) that are not well constrained by data but may be sensitive for the model

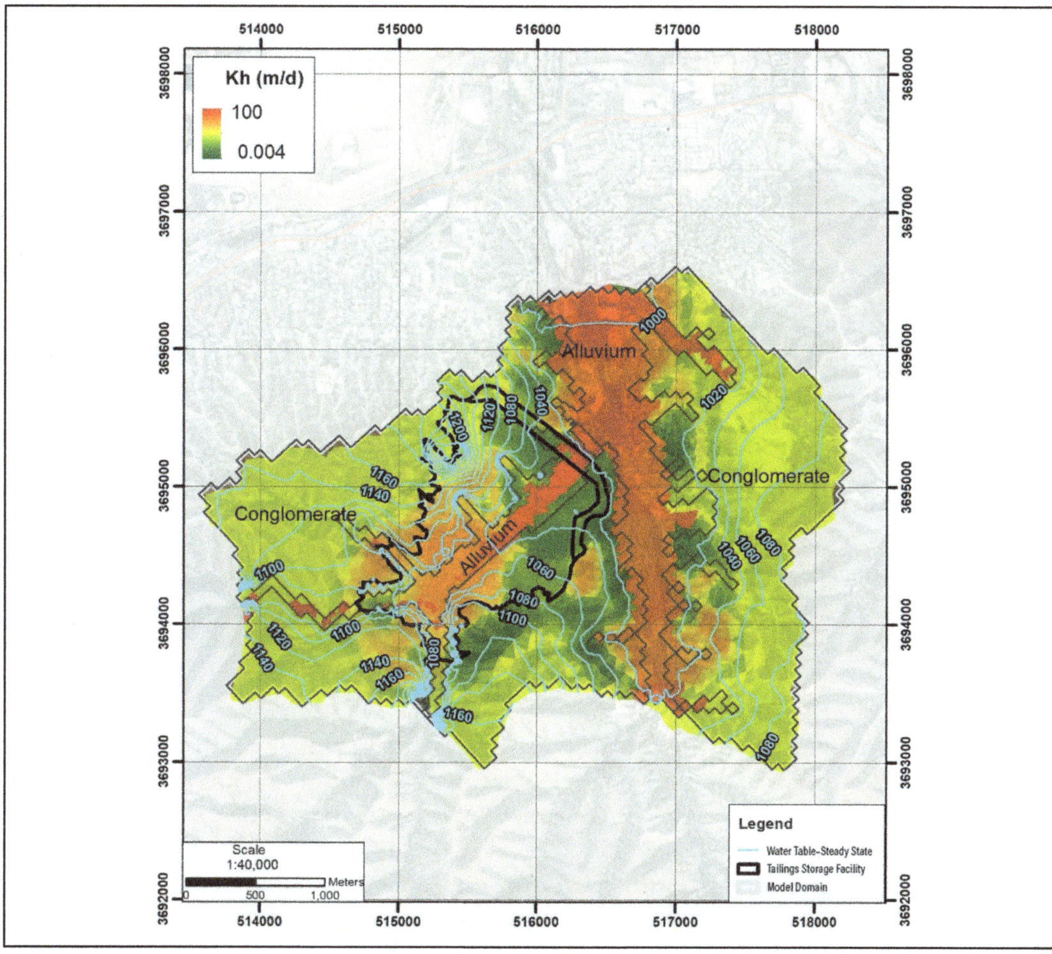

Courtesy of BHP

FIGURE 2 Example of a continuous field of simulated hydraulic conductivity values in model layer 3 from a single model realization

predictions are varied. The details of these methods are beyond the scope of this case study, but two different methods were used to explore numerous realizations of the model, including null-space Monte Carlo (Tonkin and Doherty 2009) and iterative ensemble smoother (White 2018). These techniques allow for exploration of how similarly calibrated models generate a range of groundwater levels at different simulated locations in the embankment area in response to a large storm event (Figure 3).

Uncertainty analysis results produce greater ranges of groundwater head responses at locations where monitoring is not available (distal to observation piezometers) relative to monitored locations. The simulated variability at these distal, non-monitored locations exhibits variations in head levels greater than 7.5 m, which could potentially have implications for slope stability. By comparison, head ranges were more constrained at observation points and were generally within 3 m of the measured values, with the ensemble of results enveloping the measured data as shown in Figure 3.

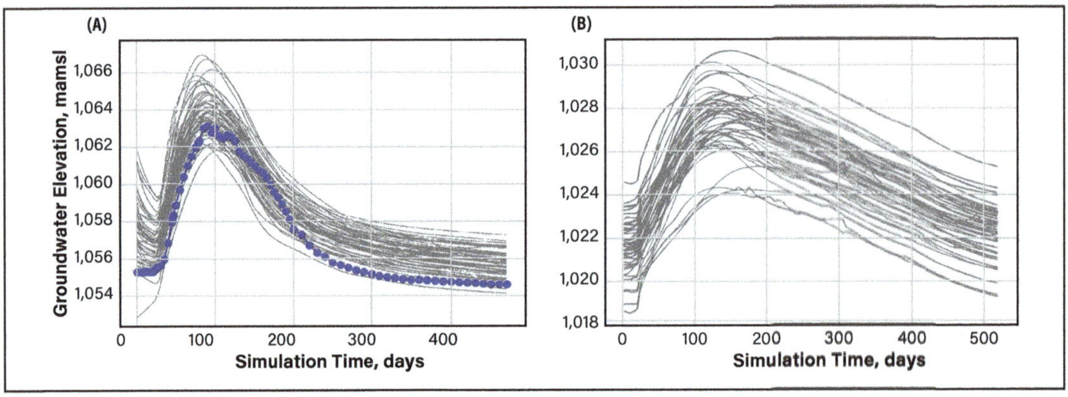

Courtesy of BHP

FIGURE 3 (A) Example of groundwater head results at a calibration point (measured data shown in blue, various realizations in black) and (B) example of groundwater heads at a point only 91.44 m from an observation piezometer. Results shown are from the iterative ensemble smoothing technique.

Transient responses across the site caused by climatic conditions or other transient boundary conditions may be highly variable. In general, much of the uncertainty in models is related to the effects of hydraulic stresses; for example, larger recharge pulses may induce higher uncertainty in model predictions. However, these transient responses are often the prediction of interest as they can be triggers for slope failures.

Another important outcome of the analysis is that model predictions are highly sensitive to the conceptual model underpinning the framework of the numerical groundwater. Tested alternative conceptual models, as well as the structural decision to represent hydraulic properties as continuous parameter fields, led to wider ranges in groundwater heads than the uncertainty analysis itself. This reinforces the need to consider valid alternative conceptual models that can be supported by available data in the modeling process.

Insights into the variability in head responses to hydraulic stresses, including, for example, seepage fluxes from the TSF during operations, will improve a tailings engineer's understanding of the range of likely responses. Such forecasts can be qualitatively or potentially quantitatively considered in the design process, where a range of predicted model values, or the associated probability distribution function itself, can be applied in the geotechnical analysis. "Will a threshold pore pressure be exceeded at a given location?" is the type of question that groundwater models are better at answering, rather than predicting precise values (Doherty and Moore 2020). For example, if the threshold pore pressure value for an unacceptable factor of safety is 1,033 mamsl (meters above mean sea level) at the monitoring location in question, the results in Figure 3B demonstrate that this could be reasonably refuted as a likely possibility based on the evaluated scenario.

Providing tailings engineers with more comprehensive groundwater model forecasts identifies which scenarios are well constrained and which ones are poorly constrained. Such approaches will equip tailings engineers and decision-makers with an improved understanding of potential hydrogeologic risks, and, by extension, actions or design changes that can reduce those risks. ■

REFERENCES

ASTM D5611-94. 2008. *Standard Guide for Conducting a Sensitivity Analysis for a Groundwater Flow Model Application*. West Conshohocken, PA: ASTM International.

Doherty, J., and Moore, C. 2020. Decision support modeling: Data assimilation, uncertainty quantification, and strategic abstraction. *Groundwater* 58(3):327–337.

Doherty, J.E., Hunt, R.J., and Tonkin, M.J. 2010. *Approaches to Highly Parameterized Inversion: A Guide to Using Pest for Model-Parameter and Predictive-Uncertainty Analysis*. Scientific Investigations Report 2010–5211. Reston, VA: U.S. Geological Survey.

Tonkin, M., and Doherty, J., 2009. Calibration-constrained Monte Carlo analysis of highly parameterized models using subspace techniques. *Water Resources Research* 45(12). doi:10.1029/2007WR006678. Accessed March 2021.

Watermark Numerical Computing. 2021. *PEST Model-Independent Parameter Estimation*, 7th ed. https://pesthome page.org/documentation. Accessed January 2021.

White, J.T. 2018. A model-independent iterative ensemble smoother for efficient history-matching and uncertainty quantification in very high dimensions. *Environmental Modelling and Software* 109 (November): 191–201.

Using a Mobile and Enterprise Geographic Information System for Improved Inspection Documentation and Reporting

Michael Wallace
Freeport-McMoRan

Mobile geographic information system (GIS) applications have immensely improved in the past few years. Paper checklists, survey forms, digital cameras, and writing in waterproof notebooks have been replaced with a smartphone or tablet and mobile applications. Avoiding illegible text, rain-smeared ink, and file cabinets with stacks of old checklists and forms are some of the many advantages to moving away from pen and paper. Today, a mobile device, a built-for-purpose application, and a web portal suffice to record and store digital survey data. Seeing these trends emerging, the corporate Tailings, Crushed Leach and Water (TCLW) group at Freeport-McMoRan (FCX) began an initiative to migrate from paper to digital forms. The TCLW group looked for a system with the following requirements:

- Mobile application that records location, photos, and videos and saves them linked with contextual information that:
  - Has a short development time
  - Is supported on different mobile platforms
  - Has secure database storage
  - Offers flexible options for data visualization, reporting, and so forth
  - Includes language support (Bahasa Indonesian and Spanish)

A form-based mobile GIS application best fit the preceding requirements. Figure 1 is a diagram detailing the architecture currently used to acquire, store, analyze, and visualize geospatial data.

The mobile side is simple with single pipes to the servers and database. On the back end, multiple applications can be connected to the data. Users can log into a web app or map to visualize, report on, and download the data for further analysis.

As an example, operators who maintain tailings discharge points such as spigots, mobile cyclones (crane or sled mounted), and tailings delivery lines previously used the form in Figure 2A to inspect each deposition location on a regular frequency. An operator visited each deposition location and filled in the paper form every 2 hours. This created a large stack of forms that required time to review.

The digital inspection form (Figure 2B) records all the same information as the paper form and automatically tracks the operator's location, date, and time. The form is accessed and filled out using a ruggedized tablet that is shared among the operators. The operators pick up

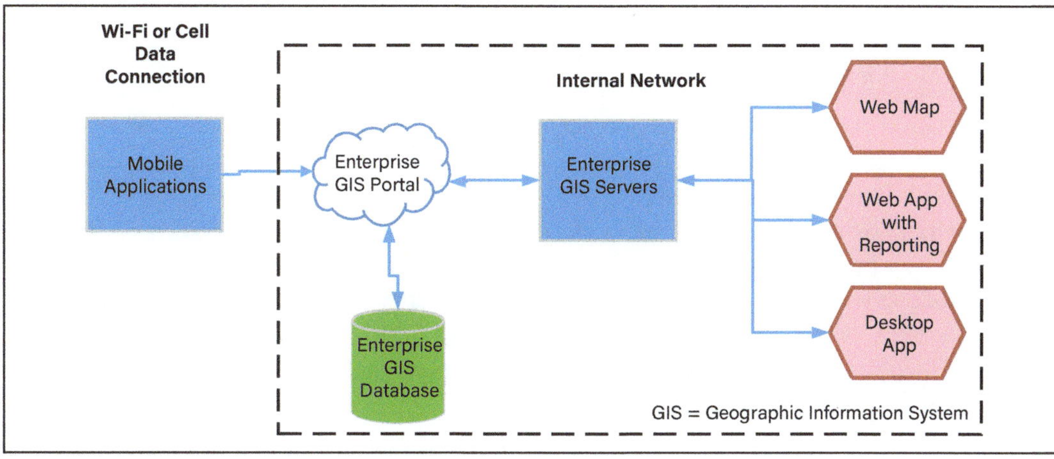

Courtesy of Freeport-McMoRan

FIGURE 1 GIS architecture connecting software, servers, and mobile devices

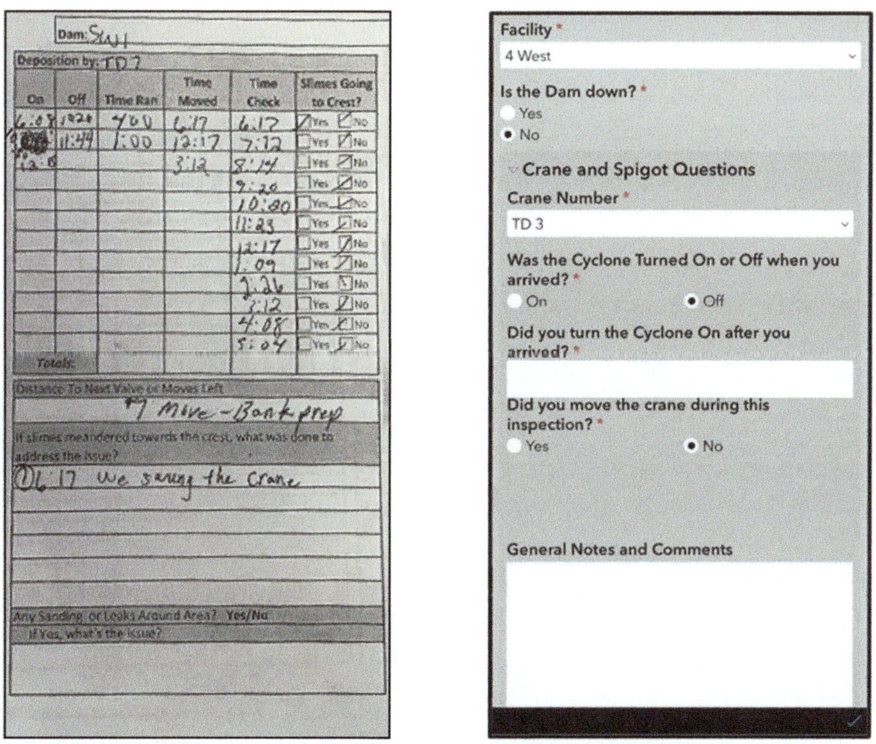

(A) Paper inspection form **(B) Digital inspection form**

Courtesy of Freeport-McMoRan

FIGURE 2 Forms used by operators to monitor deposition

the tablet at the beginning of their shift, complete their observations throughout the day, and then return the tablet to a supervisor. At this point, the data are synced via Wi-Fi to the web GIS portal for review and archival. This allows for a quick query for the most recent observations and simple download to a spreadsheet for review. The portal also enables users to create a simple report, set up email notifications, or check statistics for each question to understand the frequency of specific observations.

In addition to the site-specific example presented earlier, FCX also uses mobile forms to help the sites provide consistent documentation of facility health throughout the month. TCLW developed work instructions for performing inspections and action item tracking. These include the use of a mobile form and web GIS application for documenting inspection findings and reporting these findings routinely and tracking maintenance items over time. Site tailings engineers at all of the active FCX North American operations are using this application.

The form has four key questions that the site engineers answer from a predefined list of terms—facility, component, observation, and severity.

1. What facility are you inspecting?
   - A list of facility names for the mine site is provided.
2. What part of the embankment are you looking at?
   - The list of answers includes crest, toe, slope, abutment, access road, and so forth.

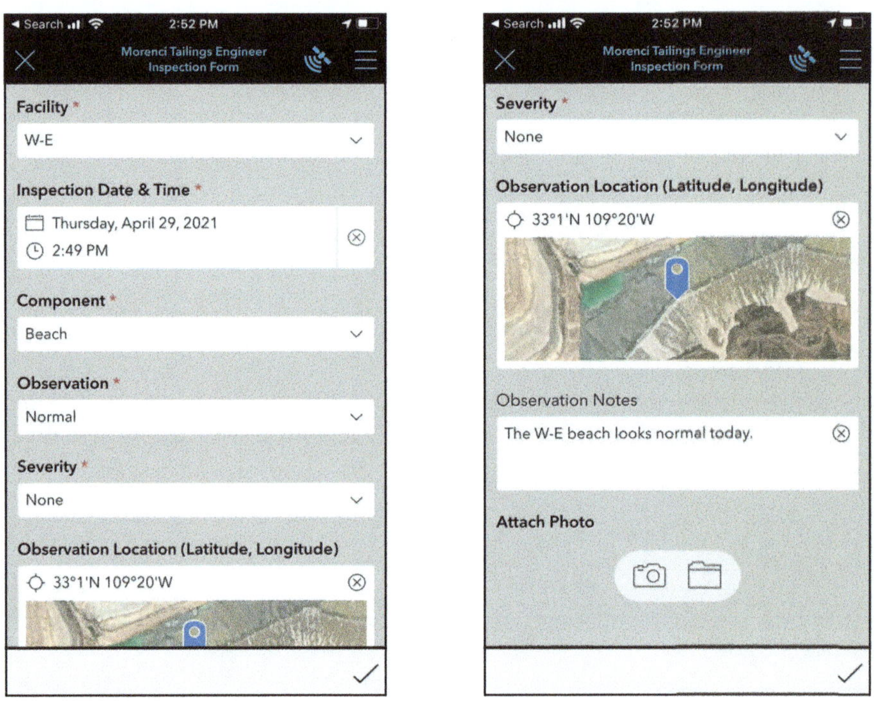

(A) Top of form **(B) Bottom of form**

Courtesy of Freeport-McMoRan

FIGURE 3 Tailings engineer inspection form

3. What are you observing related to this component?
   - The list of answers includes normal (default answer), cracking, scouring, undercutting, and so forth.
4. Does the observation require further action?
   - Answers include none (default), monitor, maintenance, and emergency action plan.

The remaining prompts include notes, pictures, and action item tracking. Standard fields, such as location, time and date, and the site, are auto-populated. Figure 3 shows the current digital form.

A web app was also developed to provide a user-friendly environment to review, edit, visualize, and report on the observation data. Several custom workflows were developed to quickly symbolize, filter, and generate reports.

Figure 4 shows a page from the observation report. The report includes a feature map, observation information, and an attached photo. It is very easy for any site engineer or corporate engineer to locate the data they are looking for and create an observation report.

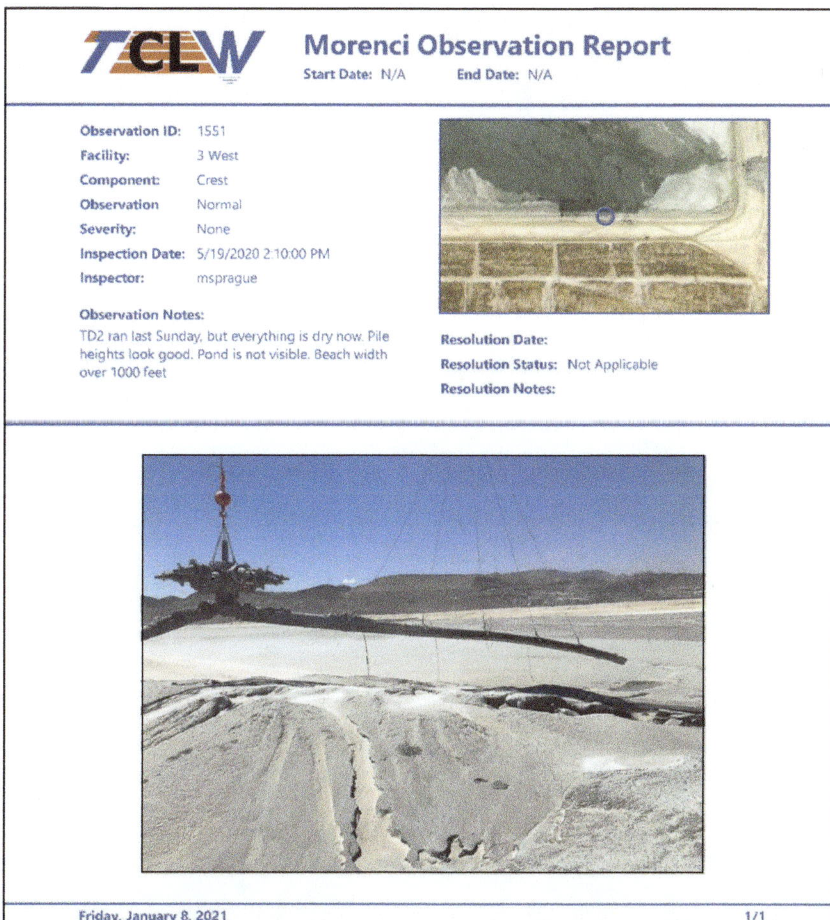

Morenci Observation Report

Start Date: N/A End Date: N/A

Observation ID: 1551
Facility: 3 West
Component: Crest
Observation Normal
Severity: None
Inspection Date: 5/19/2020 2:10:00 PM
Inspector: msprague

Observation Notes:
TD2 ran last Sunday, but everything is dry now. Pile heights look good. Pond is not visible. Beach width over 1000 feet

Resolution Date:
Resolution Status: Not Applicable
Resolution Notes:

Friday, January 8, 2021 1/1

Courtesy of Freeport-McMoRan

FIGURE 4 Monthly observation report page

Implementation and adoption of the system required planning, input, and collaboration between many stakeholders. As with any new tool, training is paramount and critical to adoption. The operators in the preceding example were trained by an engineer and worked through a transition period to familiarize themselves with the system. The tailings engineer inspection form was gradually rolled out by testing at one site, improving the system, and then releasing to the other mine sites through web meetings and on-site demonstrations.

Mobile forms allow engineers and operators to easily collect valuable information about current conditions on the FCX tailings storage facilities (TSFs). An enterprise GIS system supports a variety of workflows to visualize, report on, and analyze inspection data. Engineers and management can choose from their preferred application to digest the data such as computer-aided design, spreadsheets, ArcGIS apps, or the custom web app. Information collected informs engineers and management about facility health and broader TSF management activities such as deposition planning, maintenance needs, and assessing facility performance. Mobile GIS forms and an enterprise web GIS portal are valuable tools that allow increased collaboration and documentation within FCX's tailings management system. ■

Utilizing High-Frequency Satellite Imagery to Improve Water Balance Model Calibration and Forecasting

Brant Whiting
Freeport-McMoRan

Because of many factors, water balance models designed for tailings storage facilities (TSFs) often lack frequent measurements of reclaim water pond volume. It is common that these measurements may be conducted on an annual or quarterly time frame, but more frequent measurements can more fully capture important fluctuations in pond volume and surface area because of operational activities and climate-related events. Higher-frequency measurements also may reduce uncertainty in model calibration and results when, for example, a model is calibrated by month, but measurements of pond volume (and surface area) are quarterly or annually. Not only does the calibration of a water model critically depend on the volume measurement to make a complete accounting of all water in the system, but the low frequency of the measurement may not fully account for the applicability of the model results for periods in between.

This case study shows that by including once every 10-day measurements of TSF reclaim pond surface area using satellite imagery, the number of calibration points in the model are increased, thus producing higher confidence in model results on a weekly and monthly basis. This is because the higher-frequency calibration points assist in overcoming deficiencies in the area-to-volume relationship between surveys. This is especially relevant to tailings operators and site managers because the size and location of the pond is often of great importance to ensure a safe distance from the embankment crest.

A dynamic simulation model may be used to account for and forecast TSF water flows. A TSF water balance model must include a quantification of all inflows and outflows to the system, as well as measured volumes of the reclaim water pond—especially for facilities where the typical change in pond storage volume is equal to or greater than other flows in the system. A flow schematic of the TSF water system can aid in assuring that all inputs and outputs are identified.

Because of measurement error and uncertainty in the parameters used to calculate the inflows and outflows, it is necessary to also measure the reclaim pond volume at specified frequencies and include this volume in the TSF mass balance equation to obtain a complete accounting of the water balance.

$$\sum \text{inflows} - \sum \text{outflows} = \text{D pond storage}$$

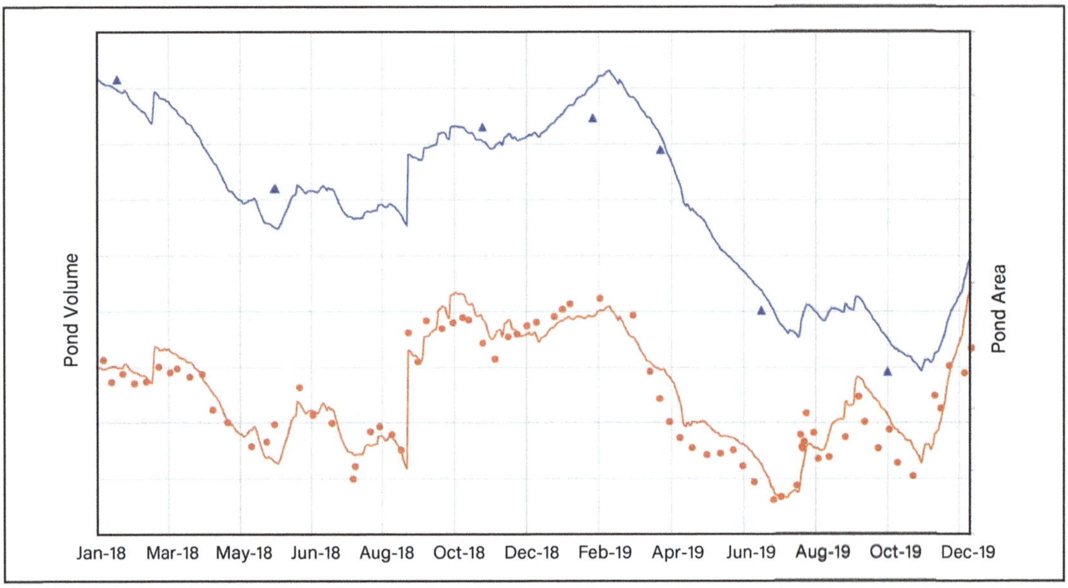

Courtesy of Freeport-McMoRan

FIGURE 1 Tailings storage facility reclaim pond volume and surface area with model simulated and measured values. The large spikes of pond volume and area in January 2018, October 2018, and August 2019 are caused by precipitation events (not shown).

TSF pond volume measurement is typically carried out using bathymetric survey techniques that measure and record the depth of clear water and corresponding latitude/longitude coordinates to characterize the deposited tailings surface geometry under the water. These surveys are important to complete, as they not only provide a direct measurement of the water pond volume and surface area, but the area-to-volume relationship also can be derived from the data. When combined with the topography above the water level, a complete rating curve can be used to simulate the changing pond area throughout the model simulation. Unfortunately, bathymetric surveys using crewed boats or remote-controlled survey boats face many challenges that can limit the number of measurements made during a given period, including inclement weather, access to the entire water area, staffing and operational constraints, equipment failure, access to water's edge, site safety standards, and floating debris. In addition to the challenges in completing bathymetric surveys, water depth measurements may be inaccurate because of signal attenuation in water with high concentrations of suspended sediment, and areas that cannot be accessed by survey boats may require interpolation of portions of the data.

Fortunately, infrequent bathymetric surveys can be supplemented by interim data collected from satellite imagery. Satellite images with up to daily frequencies can be obtained through various providers, both paid and free. The satellite images (containing the red, green, blue, and near-infrared bands) can be processed using multispectral analysis methods to determine the pond outline at which the water surface begins. This calculation is based on the normalized difference water index (Xu 2006). Using this method, pond area outlines can be obtained with high precision.

A TSF simulation model was set up and calibrated over a 24-month period. The model was calibrated by adjusting parameters such as the pan evaporation and runoff coefficients; the amount of tailings beach considered to be dry, wet, or under active deposition; and other parameters to a lesser extent. The simulation included seven combined bathymetric and topographic surveys (and their respective area–volume curves) and 74 pond surface area measurements determined using satellite imagery and multispectral analysis.

Figure 1 shows the time history of the simulated and measured pond volumes. For the seven paired volume measurements, the root-mean-square error (RMSE) was 21% of the average measured value. The simulated pond volume vaguely approximates the measured volumes, but it is unknown how accurately the model simulates between the measured pond volumes.

The figure also shows the time history of the simulated pond areas with an overlay of the measured pond areas. When comparing the 74 paired pond surface area measurements to the simulated values, the RMSE was 8% of the average measured value. Although the simulated trend line tends to over- and underestimate the measured pond areas, the simulated areas closely follow the overall trend of the measured values.

Overall, the error in the model simulated values decreased from 21% to 8% by using the measured pond areas. The benefit of this type of calibration is that the weekly and monthly values for consumption that are simulated by the model also are believed to be more accurate, and when paired with specific operational and climate information can provide operators and managers with an improved understanding of how their practices impact TSF water performance. ■

REFERENCE

Xu, H. 2006. Modification of normalised difference water index (NDWI) to enhance open water features in remotely sensed imagery. *International Journal of Remote Sensing* 27(14):3025–3033. doi:10.1080/01431160600589179. Accessed March 2021.

Use of Reclaim Water Turret Intake in a Cold Climate at the Porcupine Mine, Canada

Shaun Banks
Newmont Porcupine

Newmont Porcupine is a gold mining and milling operation located in Porcupine, Ontario, Canada, as shown in Figure 1. The climate in Porcupine averages 3 m of snow yearly with daily average freezing temperatures from November to March, and extreme temperatures as low as –45°C. The active tailings management area, 6TMA, is approximately 590 ha (hectares) in area and 12.5 km in circumference. Tailings deposition at the facility is done by the end spill and spigot method at densities ranging from 32% to 40% solids. Process water for milling operations is nearly 100% reclaimed from the reclaim water pond at 6TMA.

The original reclaim water system, as shown in Figure 2, included four vertical turbine pumps drawing water from a well at the reclaim water pond, fed from an inverted intake pipe at the bottom of the reclaim water pond. The intake has an underwater decant structure with stop logs to prevent tailings sediment from entering the reclaim water intake. The well, reclaim pumps, and effluent treatment plant feed pump were housed inside a concrete block building with concrete floor.

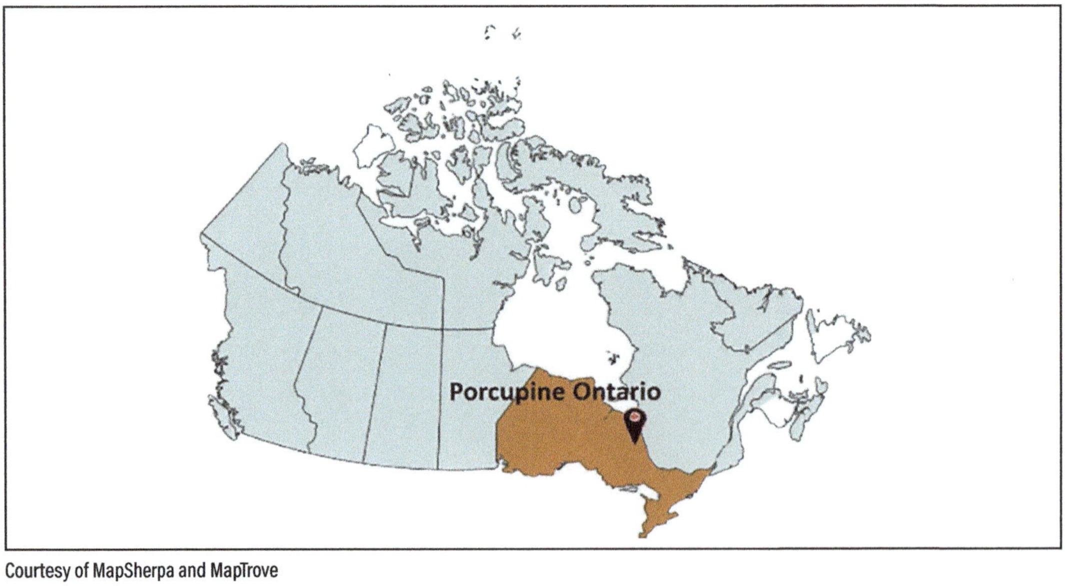

Courtesy of MapSherpa and MapTrove

FIGURE 1 Location of Newmont Porcupine, Canada

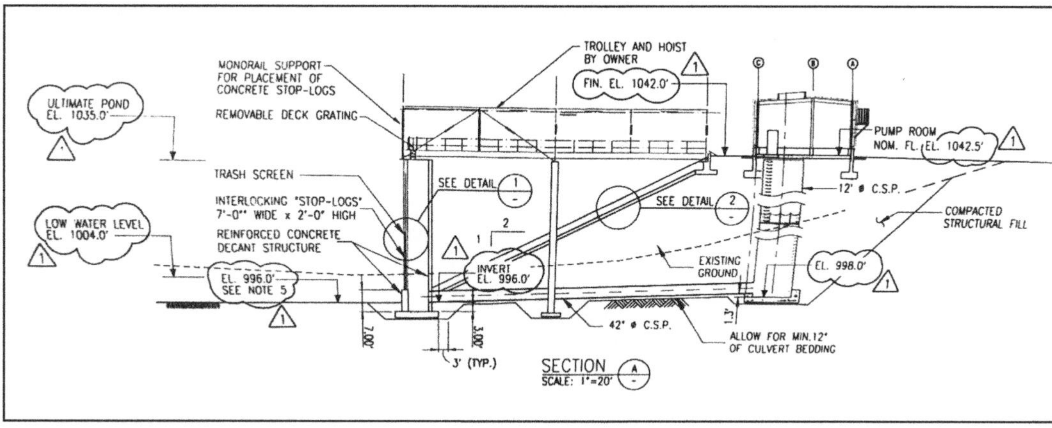

Courtesy of Newmont

FIGURE 2 Original reclaim bottom-draw inverted intake and well

As the embankment raises progress, the reclaim pumphouse is required to be raised periodically and has been raised five times. These raises are quite involved and costly. In addition, the tailings management plan involves transitioning the reclaim water pond to the central area of 6TMA, involving depositing tailings in the reclaim pond and creating beaches along the embankment. This deposition in the pond will impact the reclaim water intake so transition to a top-draw intake system, with the ability to be repositioned was required. Mining operations in cold climates in the area generally use barges with short vertical turbine pumps, anchored to shore, with de-icing using agitation or aeration.

A study was undertaken to develop options for installation of a barge. Various examples of barge installations were reviewed. Barges in warmer climates can be open to the elements, and more economical. For northern climates dealing with freezing conditions, considerations are required to prevent freezing around the barge, freezing of standby equipment, maintenance of the system in the winter, and access to the barge. The resulting concept involved a heated enclosure with lifting beams, access gallery, and de-icing, leading to a robust barge and costly option, with some practicality issues remaining.

It was suggested to Newmont Porcupine to explore the option of using a turret intake with a pump on shore as used in Australia. The advantages of the turret intake include requiring only 1 m depth of water as well as flexibility to be relocated easily. Figure 3 shows a turret intake installation in Australia.

Contact was made with the company that designs and supplies turret intakes for Newmont Boddington, Australia. There was a high level of confidence that the turret intake system would work in our application, and the challenge was to ensure that it would not be impacted by freezing conditions. Criteria in the design included de-icing, maintenance of the pumps in the winter, and an easily movable system. The turret supplier began designing a de-icing system while Newmont Porcupine worked with a multi-disciplinary consultant for engineering of the piping, electrical, and enclosure. A de-icing pipe with nozzles was designed on each side of the suction pipe and around the turret intake. The first idea was to recirculate a small amount of water from the discharge. To maximize the water supplied from the pumps, an alternative was considered to use a blower to supply air to the de-icing ring; however, the volume of air required a large blower as well as heating for the intake air. This significantly increased the

Courtesy of Turret Engineering Pty. Ltd.

FIGURE 3 Turret intake installation at Newmont Boddington, Australia

Photo by S. Banks

FIGURE 4 Turret intake with cone and de-icing piping

complexity, required power, and cost. The final de-icing design reverted to the original idea of using re-circulated water from the discharge. A cone was added to the 3-m diameter turret intake to prevent snow and ice build-up. A self-priming centrifugal pump is used with the turret intake. A 6-m sea container was modified with a reinforced floor, doors, electrical, ventilation, and heating to house the pump. Figures 4 through 6 show the components of the turret system.

A trial was conducted during the winter of 2022–2023 by installing one turret system. The turret pump discharge was connected to the existing header for the vertical turbine pump system. The one centrifugal pump was trialed during the winter in conjunction with the existing vertical turbine pumps to confirm the ability of the suction intake and pump to supply sufficient flow and the effectiveness of the de-icing system. The trial was successful and confirmed the ability of the turret pump to supply the mill, with supplement from an additional pump when required. The de-icing system worked better than expected. Figure 7 shows the effectiveness of the de-icing system, and Figure 8 shows the turret intake and de-icing system.

Photo by S. Banks

FIGURE 5 Modified sea container pumphouse

Photo by S. Banks

FIGURE 6 Self-priming centrifugal pump

Photo by S. Banks

FIGURE 7 Effectiveness of turret de-icing

Photo by S. Banks

FIGURE 8 Turret intake and de-icing

Following the success of the initial winter trial, work began to expand the turret system from one turret to three turrets. Refinements were made to the design, and a header was incorporated to tie the three pumps together into the reclaim piping system. The reclaim water system and effluent treatment plant feed has been fully transitioned to supply from the three turret pumps. Figure 9 shows the arrangement of the three-turret system.

The following functionality is incorporated into the system design:

- All three turret pumps are powered by variable frequency drives for variable flow demands.
- All three pumps can supply the mill reclaim water demand, providing redundancy.
- Two of the three pumps are piped to supply the effluent treatment plant, providing redundancy.
- One pump is able to supply normal reclaim water demands with ability to be supplemented by a second pump for increased flow requirements.
- One pump is able to supply effluent treatment plant demand.
- De-icing rings can be back-fed from the operating pump or pumps to ensure that all three intakes remain de-iced.
- Access is available to the side of each sea container to remove a pump if needed to take to the shop for repairs.
- There is flexibility to take one turret pump offline to raise or relocate while the others are operating.

Photo by S. Banks

FIGURE 9 Three-turret system

This turret system provides the following benefits over the existing fixed bottom-draw reclaim system:

- Ability to easily relocate with pond transition
- Can co-locate turrets or can locate them in different areas of the pond (once power feed is routed where required)
- Ability to deposit tailings and beach in the area of the existing fixed reclaim water intake that is now redundant
- Can reduce the amount of water contained in the reclaim pond due to dead zone created by the existing stop log intake
- Ability to pump in shallow water, as little as 0.4 m depth
- Capacity to extend intake further into pond
- Easier maintenance of the centrifugal pumps compared to vertical turbine pumps with extended shafts

The innovative design of a turret intake developed in Australia has been further adapted for suitability in cold climates, such as Porcupine, to withstand the winter conditions, providing operational flexibility and benefits. This is an example of a "Down Under" design adapted for the "Great White North." ■

Importance of Cross-Functional Communication and Collaboration in Developing a Successful Cemented Paste Backfill Product

Ryan L. Veenstra
Newmont Corporation

This case study highlights the importance of clear lines of communication and consultation between project engineering teams and on-site technical services teams, particularly when making alterations to the design of tailings products.

The mining operation in this case study (Newmont's Tanami operations) was initially set up to manufacture a cemented paste backfill (CPB) by harvesting tailings from a surface tailings facility. These tailings were a by-product of processing near-surface ore obtained from open pit mines. Due to the weathering and oxidation of this ore, the tailings were very fine with high amounts of clay (solid, dark red line in Figure 1). The operation, at the time, chose to proceed with harvested tailings for various reasons (cost, closure requirements, etc.), but a major consideration was the 45-km distance between the mill and the backfill plant (located at the operating underground [UG] mine), as the mill was built where the historic open pit mines

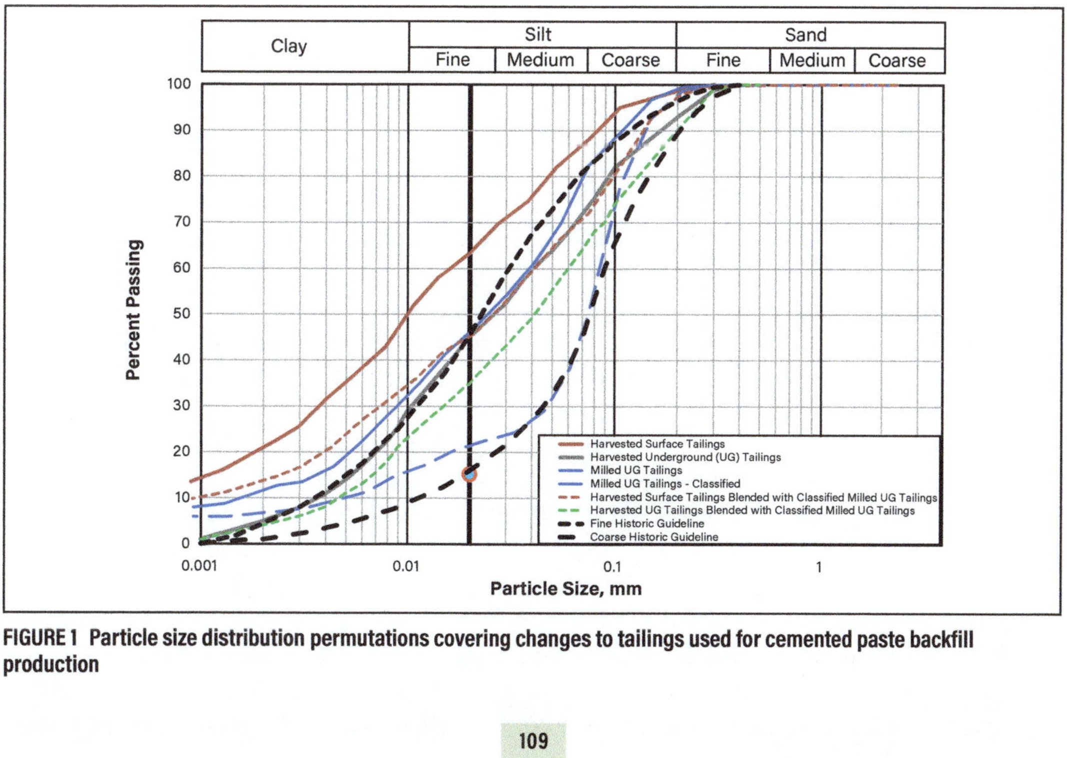

FIGURE 1 Particle size distribution permutations covering changes to tailings used for cemented paste backfill production

were located. These tailings required the CPB to manufacture at relatively low bulk densities (~60%–62% solids by mass), which necessitated more binder to achieve the needed strengths (as compared to a backfill made from *fresh* UG tailings). To deliver the densest CPB possible (to keep the binder costs as low as possible), the reticulation system was built using 200-mm piping. As the harvested tailings sources approached depletion, the operation decided to switch to the operating mill's tailings stream.

The "15 passing 20 rule" means that CPB requires 15% of the tailings to be finer than 20 μm, which is an old industry design guideline for defining the difference between a hydraulic fill (HF) and CPB. This guideline was linked to the different transportation environments (turbulent flow for HFs and laminar flow for CPB). The rule was likely adopted from the concrete industry (like many of the old design guidelines in the backfill industry), where most pumpable concrete mixes require about 15% fines, which is made up of the added cement (Stone 2014).

In the figure, the 20-μm line has been thickened and the 15-20 guideline is shown as the blue dot. The dashed black curves show a further modification to this guideline: Particle size distributions (PSDs) coarser than the *coarse historic guideline* were sufficient for HF; PSDs that were finer than the *fine historic guideline* were marginal for CPB and cannot be used for HF; and PSDs falling between the guidelines were sufficient for CPB but required classification for HF. Although this rule and its modifications can be used as a starting point, further test work needs to be completed looking at the mineralogical, chemical, and rheological properties.

However, the operation decided to design the system using the 15–20 guideline and did not conduct any specific backfill-related test work. The design involved classifying the full-stream mill tailings (solid blue line in the figure) using a typical hydrocyclone bank (the classified PSD curve is shown as the dashed blue line in the figure). The coarse material was then sent to a belt vacuum filter for further dewatering, so that it could be trucked the 45 km to the backfill plant. The system was relatively inexpensive (compared to the cost of filtering technology for a finer-grained material) and worked well. However, it did not produce a product that could be made into a CPB, despite meeting the 15–20 guideline. The classification had poorly graded the tailings (as is typical with hydrocyclones), producing a product that behaved like an HF. This, combined with mineralogical and chemical differences, did not allow the classified tailings to be used to make a CPB, particularly given that the UG mine's reticulation network had large-diameter pipes.

To create a CPB using this setup, the operation attempted to run finer material over the belt filter, but the filter was unable to sufficiently dewater this material (both because of the decreased feed density and the additional fines). To allow backfilling to continue, a tailings blend was created by combining filter belt product with the remaining harvested oxide tailings. Various blends were trialed, eventually arriving at a 50:50 mix (the dark red dashed line in the figure). This ran well until the oxide material was depleted and a new 50:50 mix was generated, blending harvested mill tailings (UG tails that were deposited into a tailings facility and then harvested) (gray line in the figure) and filter belt material (dashed green line in the figure). This material was usable but was significantly more difficult to control within the reticulation network.

While these harvested blends allowed the operation to backfill (and kept the operation producing), it was not economically viable in the long run as the operation had to run the harvesting operation, the classification/dewatering circuit, as well as the blending operation. Although the harvest tailings usage was reduced at this rate, the available sources were approaching depletion. To move to full mill tailings, the operation decided to upgrade the existing filter belt to dewater the full-stream mill tailings (solid blue line). These upgrades are currently underway, and while the upgrades are expected to be successful and allow the finer material to be dewatered, the throughput of the filter belt is also expected to decrease, likely requiring the purchase and installation of an additional vacuum belt filter. Adopting this new dewatering setup also made the hydrocyclone bank redundant.

In total, it took around 3 years for the operation to work through this process, which required the technical services team to continually adapt their designs to allow the operation to keep safely producing. Ultimately, there was a lack of communication about the project's end goal. The original design material created was never going to be feasible for producing backfill at the operation's UG mine. However, the design of this material was derived from what was attractive on the material processing side and not what was required to actually create a usable backfill product. This lack of understanding led to increased operational costs and, ultimately, significantly more capital cost in order to provide a usable tailings product. ■

REFERENCE

Stone, D. 2014. The evolution of paste for backfill. In *Mine Fill 2014: Proceedings of the 11th International Symposium on Mining with Backfill.* Edited by Y. Potvin and T. Grice. Perth: Australian Centre for Geomechanics. pp. 31–38.

PART III

Closure and Reclamation

Approach to Safe Closure of Tailings Storage Facilities

Brett Byler
Newmont Corporation

Jonathan H. Gillen
Newmont Corporation

Closure planning is a key component of integrated mine planning through all phases of the tailings storage facility (TSF) life cycle. TSFs should be planned and designed from the outset with the goal of developing a final closure plan that meets closure and post-closure objectives, achieves stable landform(s), and meets the definition of *safe closure.* Newmont's Aunor A TSF in Ontario, Canada, shown in Figure 1, is one such facility that has been reclaimed in an effort to achieve a stable landform status. The Global Industry Standard on Tailings Management (GISTM) (GTR 2020) defines safe closure as:

> A closed tailings facility that does not pose ongoing material risks to people or the environment which has been confirmed by an ITRB [independent tailings review board] or senior independent technical reviewer and signed off by the Accountable Executive.

Courtesy of Newmont Corporation

FIGURE 1 Aunor A Tailings Storage Facility, Ontario

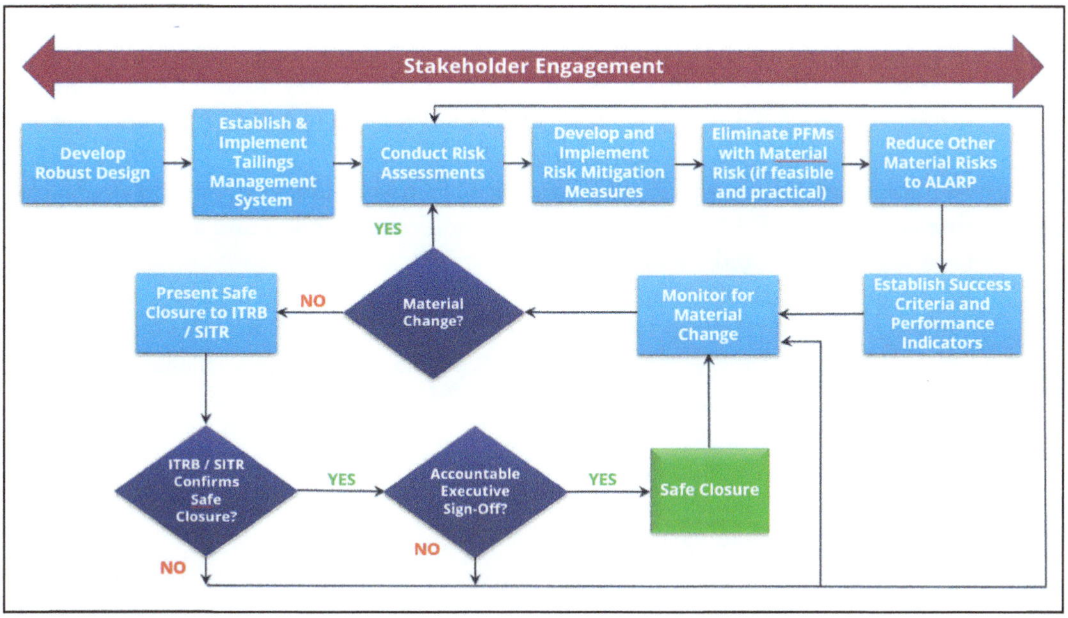

Courtesy of Newmont

FIGURE 2 Proposed road map to safe closure (ALARP = as low as reasonably practicable; ITRB = independent tailings review board; PFM = potential failure mode; SITR = senior independent technical reviewer)

Based on this definition, achieving safe closure inherently requires a risk-based, and thus facility-specific, closure planning strategy. Every TSF is unique and has unique risks and challenges with site-specific conditions influencing the ability to achieve safe closure. These site-specific conditions may include the social context and social impact; current closure status, including whether the facility is new or already-constructed; legal status; regulatory environment; mine plan; reclamation plan; seismicity; climate; geology; geomorphology; watershed characteristics; hydrogeology; and material properties (tailings, waste, and construction materials).

Newmont has developed an internal safe closure guidance document that provides a framework, or "road map," to safe closure as illustrated in Figure 2. This framework accounts for site-specific considerations and is implemented across all stages of the TSF life cycle, including facility siting, design, construction, operation, and closure. Furthermore, engagement with key stakeholders, including local communities, Indigenous Peoples, and regulators, must occur throughout the life cycle of the TSF. Key steps for the safe closure framework include:

1. **Begin with the end in mind.** Develop a robust design for the TSF that demonstrates the feasibility of safe closure. Robust designs should include multiple and redundant controls that mitigate potential failure modes with material risks. Some examples of robust design include:
   - Adopt design criteria for seismic and hydrologic loading consistent with Extreme Consequence Classification facilities per the GISTM (GTR 2020).
   - Identify and address brittle failure modes with conservative design criteria, independent of trigger mechanisms, to minimize their impact on the performance of the TSF.
   - Eliminate potential failure modes that could result in a catastrophic release of tailings and water to the extent practical and feasible.

- Employ integrated mine planning to buttress the TSF with deformation-tolerant waste rock dumps or heap leach facilities.
- Strategically densify tailings with the objective of achieving dilatant conditions throughout the tailings deposit (e.g., for filtered tailings facilities) or in critical locations such as zones near containment structures.
- Minimize or eliminate water stored on the TSF. This includes designing the TSF closure landform so that it does not store water and cannot collect and accumulate water that could initiate and propagate an overtopping failure, potentially resulting in an uncontrolled release of tailings.
- Promote unsaturated tailings within the TSF. For example, include over-liner drainage systems for lined TSFs to promote tailings consolidation and unsaturated conditions. Gravity-drainage is usually preferred, if possible, as long-term pumping is typically not possible or desirable.
- Implement tailings dewatering technologies, such as thickened, high-density thickened and filtered tailings during the operation of the facility, or during closure to support landform grading activities.
- Commingle tailings and waste rock to improve material shear strength characteristics, minimize acid production and metals leaching, facilitate rapid and progressive reclamation, and reduce the mine's waste storage footprint.
- Locate TSF closure spillways on dam abutments or elsewhere on a competent rock foundation. Location of spillways on the TSF embankment should be avoided when possible. Spillway location and design considerations should include conveying water to avoid potential impacts to the TSF embankments, downstream communities, and the environment. The spillway should be located to direct flows in a controlled manner to native stream beds, where possible.
- Evaluate and implement strategies to prevent or limit impacts from acid generation and metals leaching. Strategies may include avoiding the processing of particularly reactive ore types; tailings desulfurization; segregation or selective handling (e.g., encapsulation or layering) of potentially acid generating (PAG) and non-acid generating (NAG) materials; commingling of tailings and waste rock to reduce oxygen ingress or flow rates; alkaline amendment of tailings; and design and construction of engineered barrier systems (i.e., liners and covers). If water covers are considered as a strategy to limit oxidation of tailings, the benefits should be compared to the risks of maintaining water behind embankments. Maintaining a water cover may increase both the likelihood and consequences of an embankment failure.
- Mitigate or eliminate long-term seepage and contamination of groundwater and surface water. Contaminant transport should be evaluated in the design phase, and mitigation measures should be adopted to limit adverse impacts to the environment and downstream communities. Groundwater monitoring and periodic calibration of a groundwater model should be conducted to allow for modifications to groundwater monitoring and mitigation programs. Baseline groundwater levels and water quality data should be collected prior to construction and operation of the TSF.
- Avoid or limit reliance on manufactured materials, such as geomembranes and geotextiles, which have the potential to degrade over time.

- Identify construction and closure material requirements, including material availability and required physical and geochemical characteristics. These materials should be sourced and stockpiled throughout the mine life to avoid contamination with unsuitable material. A closure material balance should be developed and periodically checked.

2. **Establish and implement a tailings management system (TMS).** The TMS consists of processes and procedures for design and management of the TSF. Following the Global Tailings Review (GTR; 2020), key elements of the TMS include establishing performance objectives; change management; identifying and securing adequate resources, such as experienced and qualified personnel, equipment, scheduling, data, documentation, and financial resources; conducting robust performance evaluations and risk assessments; establishing and implementing controls for risk management; auditing and reviewing for continuous improvement; implementing a management system with clear accountabilities and responsibilities; and preparing and implementing an effective operations, maintenance, and surveillance (OMS) manual and emergency response plan (ERP). A robust TMS provides the guardrails to keep construction, operation, and closure of the tailings facility on the path to safe closure.

3. **Conduct risk assessments of the TSF throughout the TSF life cycle.** Risk assessments should be conducted to:
   - Identify material risks to people and the environment through conducting performance reviews and dam safety reviews, effective surveillance, and periodic risk assessments.
   - Document material geotechnical, hydrotechnical, geochemical, and hydrogeochemical risks to people or the environment. Examples of material risks to people and the environment include:
     - Geotechnical and hydrotechnical potential failure modes resulting in catastrophic failure (e.g., overtopping, internal erosion, foundation failure) of the TSF embankment.
     - Noncompliant drainage from the TSF to surface water or groundwater points of compliance.
     - Unacceptable fugitive dusting from the TSF.

4. **Develop and implement risk mitigation action plans.** Key steps are:
   - Develop risk mitigation plans based on the results of risk assessments, performance reviews, and dam safety reviews to reduce the likelihood and/or consequence of potential failure modes and material risks to people or the environment.
   - Implement risk mitigation plans to reduce material risks to people and the environment to as low as reasonably practicable (ALARP).
   - If feasible and practical, eliminate potential failure modes with material risks to people and the environment. In many cases, it may not be feasible or practical to eliminate a potential failure mode.
   - When potential failure modes cannot be eliminated, implement risk reduction measures that reduce the likelihood and/or consequence of the potential failure mode.

5. **Establish performance indicators and monitor for change.** Performance indicators are critical metrics used to evaluate whether a facility is operating as intended and as expected. Key steps are:

- Establish performance objectives and indicators for the implemented risk mitigation actions and controls. Continue TSF surveillance to verify performance objectives are achieved and monitor for material change in expected TSF performance.
- If material changes to the TSF or performance are identified, return to Step 3. Examples of material change might include unexpected degradation of drainage from the TSF or foundation displacement or deformation exceeding the anticipated performance and/or acceptable thresholds.

6. **Demonstrate and confirm safe closure with an independent tailings review board (ITRB) or senior independent technical reviewer (SITR).** Key components of these independent reviews include:
   - Demonstrating that the activities in Steps 1 through 5 have been thoroughly implemented and completed is the key to achieving a safe closure designation. Once a TSF has been closed, risk mitigation measures have been implemented, and sufficient monitoring data have been collected to demonstrate performance objectives are being met, the documentation of these activities and mitigation measures are presented to the ITRB or SITR for review.
   - If the ITRB or SITR confirms that safe closure has been achieved, the confirmation should be documented by the ITRB or SITR.
7. **Accountable executive sign off that safe closure has been achieved.**
   - Once an ITRB or SITR has confirmed a safe closure designation, the final step is for the accountable executive, as defined by the GISTM (GTR 2020), to sign off that a safe closure designation has been achieved with written documentation.
8. **Continue to monitor the TSF for material change.**
   - Continue surveillance and maintenance of the TSF. If material change(s) are identified, return to Step 3.

Safe closure *does not* mean that governance activities, surveillance, and other risk management activities cease. As part of a responsible tailings management framework, continued effort is required to ensure that TSFs with a safe closure designation remain safely closed. The requirements for governance, surveillance and maintenance after a safe closure designation will vary and should be established based on a risk-informed process.

Governance, surveillance, and risk management activities continue, as appropriate on a site-specific basis, after demonstrating safe closure. Those activities may include:

- Maintaining an engineer of record (EoR) for the TSF after safe closure, unless consensus has been reached that the EoR role is no longer considered necessary. Elimination of the EoR role for the TSF is a risk-based decision, which should be reviewed by an ITRB or SITR; and signed off on by the accountable executive. For tailings facilities with a safe closure designation that do not have an EoR, a material change should initiate a risk assessment and consideration for the need to re-establish an EoR.
- Continuing performance reviews or dam safety inspections by the EoR or SITR. A reduced frequency for these reviews and inspections may be implemented with concurrence from the accountable executive and the EoR, with review by the ITRB or SITR.

- Continuing surveillance activities (i.e., inspections and monitoring) to verify that performance objectives and indicators are being met. The frequency of surveillance activities may be adjusted on a facility-specific basis with concurrence from the accountable executive and the EoR, with review by the ITRB or SITR.
- Continued maintenance of the TSF, including water management structures, cover systems, facility slopes, seepage management systems, and other closure elements. The frequency and type of maintenance will be facility specific.
- Monitor the TSF for material change through surveillance, maintenance, and performance reviews. Update the risk assessment if material change is identified. Periodic risk assessments should be considered on a facility-specific basis after safe closure.
- Regular ITRB or SITR meetings may not be necessary once a safe closure designation is achieved. However, a material change should initiate evaluation by the EoR, with consideration of independent review. For tailings facilities with a safe closure designation that do not have an ITRB or SITR, a material change should initiate a risk assessment and consideration for the need to re-establish an ITRB or SITR.

TSF closure is interwoven throughout the GISTM as a key aspect of each stage of the TSF life cycle. Design, construction, and operating decisions made for the TSF impact the implementation and performance of closure. To most efficiently and effectively achieve safe closure, the "begin with the end in mind" approach is essential. ■

REFERENCES

GTR (Global Tailings Review). 2020. *Global Industry Standard on Tailings Management.* https://globaltailingsreview.org/wp-content/uploads/2020/08/global-industry-standard_EN.pdf. Accessed November 2023.

Newmont Corporation. 2023. *Tailings Storage Facility (TSF) Closure Guideline.* January 20. www.newmont.com. Accessed November 2023.

Anatomy of a Tailings Reclamation Project: San Xavier Tailings Storage Facilities, Arizona, United States

Krishna Parameswaran
tfgMM Strategic Consulting

This case study is based on prior publications discussing reclamation of tailings storage facilities (TSFs) conducted at the ASARCO LLC (Asarco) Mission mine in southern Arizona in the United States (Parameswaran 2015, 2016, 2019, 2020). The objective of the case study is to highlight good practices in stakeholder engagement, collaborative partnerships, and project management oversight.

A portion of Asarco's Mission mine is located within the boundaries of the San Xavier District of the Tohono O'odham Nation. Asarco and the Nation entered into a lease agreement in 1959 to allow use of the Nation's property for copper mining and related activities.

In 2008, the Tohono O'odham Nation, San Xavier District, San Xavier Allottees Association, the United States, and Asarco finalized a comprehensive mining and reclamation plan (MARP) and established a $33 million fund for reclamation of tailings embankments (TSFs 1 through 3), waste rock deposition areas, and long-term maintenance activities. Asarco and the Nation formed a six-member panel to oversee and administer the implementation of the reclamation design and construction phases of the project.

The panel, which included a diverse team of technical personnel, also relied on independent consultants and governmental organizations to provide input and guidance. The U.S. Bureau of Indian Affairs, U.S. Bureau of Land Management, and the University of Arizona were instrumental in developing a very specialized revegetation matrix of reclamation seeds, trees, and cacti. A portion of the grass seed mix was grown locally by tribal industry to generate a grass variety truly adapted to the reclamation area.

Reclamation activities commenced in 2009 and were completed in 2012. The reclamation project scope involved capping with alluvium, rock armoring, construction of stormwater controls, and the revegetation of nearly 890 ha (hectares) of TSFs and waste rock deposition areas. The MARP incorporated sustainable desert vegetation goals while minimizing the long-term maintenance and respecting the significant cultural and religious values associated with the land. Balancing these factors into a comprehensive reclamation plan in a very arid region was challenging.

The surface of the TSFs was capped with a minimum of 0.3 m of alluvium and seeded with a mix of native plant species. Steep slopes and benches were graded to a more stable 3:1 angle and covered with 0.3 m of alluvium, to reduce stormwater infiltration, and 0.3 m of rock to minimize soil erosion.

The top, benches, and slide slopes of the TSF prior to reclamation are shown in Figures 1 and 2, respectively. A photograph of reclamation in progress is shown in Figure 3. The success of revegetation on the tops and side slopes of the reclaimed TSFs in October 2020, the eighth year after revegetation is shown in Figures 4 and 5, respectively. Recent photographs taken in September 2023 further illustrate the success of revegetation, with Figure 6 showing the top of the TSFs and Figure 7 showing the slide slopes.

Figures 4 through 7 clearly indicate the success of revegetation on the tops of the TSFs. Revegetation cover is less on the slide slopes because of the coarse rock used to armor the slide slopes for erosion control. In certain areas, where less coarse rock was used, the revegetation cover is better. The last survey, conducted in October 2020 (a previous survey was conducted in 2017), concluded that the revegetation met the required standards, that is, sustainable desert vegetation

Courtesy of ASARCO LLC

FIGURE 1 Tailings storage facilities before reclamation

Courtesy of ASARCO LLC

FIGURE 2 Benches and slide slopes of tailings storage facilities before reclamation

Courtesy of ASARCO LLC

FIGURE 3 Completed tailings tops and slide slopes prior to revegetation

Courtesy of San Xavier District and San Xavier Allottees Association

FIGURE 4 Top of reclaimed tailings storage facility looking northwest (October 2020)

Courtesy of San Xavier District and San Xavier Allottees Association

FIGURE 5 Slide slope of reclaimed tailings storage facility (October 2020)

Courtesy of San Xavier District and San Xavier Allottees Association

FIGURE 6 Top of reclaimed tailings storage facility (TSF), facing northeast with a large burrow pit in the foreground and reclaimed TSF 2 and 3 in the background (September 2023)

Courtesy of San Xavier District and San Xavier Allottees Association

FIGURE 7 Slide slope of reclaimed tailings storage facility (TSF) facing southeast with currently operational TSF 4 and East Rock Deposition Area (September 2023)

goals while minimizing the long-term maintenance and respecting the significant cultural and religious values associated with the land. Also, return of wildlife to the site has been noted. The success of the reclamation project has generated much interest from regulatory agencies and the mining industry throughout the United States and many foreign countries. ■

REFERENCES

Monisha, B., and Mckenna, C. 2021. *Vegetation Survey: Tailings and Mission North Dump, San Xavier District of the Tohono O'odham Nation*. GeoSystems Analysis Inc. for San Xavier District of the Tohono O'odham Nation, May 25. pp. 57–58.

Parameswaran, K. 2015. Mining and sustainable development: Oxymoron or Rx for a bright future? In *Proceedings of Water, Air and Land: Sustainability Issues in Mineral and Metal Extraction (WALSIM) Symposium at Conference of Metallurgists (COM) 2015*. Westmount, QC: Canadian Institute of Mining, Metallurgy and Petroleum. pp. 1–19.

Parameswaran, K. 2016. Sustainability initiatives at ASARCO LLC: A mining company. In *Perspectives in Metal Sustainability: Global Challenges, Consequences and Perspectives*. Edited by R.M. Izatt. New York: Wiley. pp. 424–452.

Parameswaran, K. 2019. Sustainable development and the primary copper industry-past, present and future. In *Proceedings of the Phil Mackey Symposium: Copper 2019: Conference of Metallurgists (COM)*. Westmount, QC: Canadian Institute of Mining, Metallurgy and Petroleum. pp. 1–20.

Parameswaran, K. 2020. Case study 1: Anatomy of a tailings reclamation project: San Xavier tailings storage facilities. In *Tailings Management Handbook: A Life-Cycle Approach*. Edited by K.F. Morrison. Englewood, CO: SME. pp. 877–878.

Closure of a Conventional Tailings Storage Facility in a Seismically Active Tropical Environment, Marlin Mine, San Marcos, Guatemala

Manuel Aparicio
Newmont Corporation

Leonardo Escott
Newmont Marlin Mine

Clint Strachan
Stantec

The Marlin mine is located approximately 120 km west-northwest of Guatemala City, in a mountainous region of western Guatemala. The area exhibits high seismic activity, influenced by strike-slip faulting and subduction zone activity from a regional fault system. Long-term (post-closure) conditions were analyzed for ground motions from the maximum credible earthquake (MCE). In the site area, the MCE is a Richter magnitude 8.9 event, which can generate an estimated peak ground acceleration in rock of 0.51*g* (or 51% of the acceleration of gravity). The regional climate is tropical, reflecting distinct wet and dry seasons with an average annual precipitation and evaporation of approximately 1,200 mm and 1,700 mm, respectively. Average maximum and minimum temperatures are 25°C and 10°C, with an average annual temperature of 18°C.

Mining and milling operations commenced at Marlin in late 2005. Open pit mining was phased out in 2012, while underground mining continued to the end of milling operations in mid-2017. Processing for precious-metal recovery used conventional milling, with a cyanide leach process and Merrill–Crowe circuit for recovery of gold and silver. Mill production averaged 4,500 t (metric tons) per day.

TAILINGS AND WATER MANAGEMENT

The tailings storage facility (TSF) was constructed in a valley downstream of the mill, which was designed and operated to store mill tailings and process water (Figure 1). Containment is provided by the underlying volcanic rock and a cross-valley dam (main dam) and extension along a ridge (saddle dam). The dams were designed as water-retaining structures, with low-permeability clay cores, and filter, drain, and rock-fill shell zones. The clay cores included key trenches excavated into the foundation, which was supplemented with grout curtains where

Courtesy of Newmont

FIGURE 1 Tailings storage facility at the end of mill operations, December 2017

necessary. The dams were constructed in stages in a downstream manner, with the maximum height of the main dam reaching approximately 88 m.

Water from the mill process, underground mine dewatering, and surface water runoff was retained in the TSF, providing mill makeup water in the dry season. Excess water in the TSF could be discharged if water-quality standards were met. Mill tailings were deposited into the TSF as slurry (at approximately 35% solids). The average tailings density in the TSF increased from 0.7 t/m^3 at start-up to approximately 1.3 t/m^3 at the mill shutdown. Excess water in the TSF resulted in subaqueous discharge in the first several years of operation, while later improvements in water management allowed for subaerial discharge and beach development.

PRE-CLOSURE MANAGEMENT

With the TSF dam crest elevations established, a spillway was designed and constructed to accommodate operating and post-closure conditions. It included a labyrinth wall at the invert, with sluice gates at the sides with different elevations to allow permitted discharge of water from the TSF during the rainy season. The spillway wall also provided accommodation for the tailings surface elevation at closure because the final volume of tailings to be discharged in the TSF was not established at the time of spillway construction.

As mine plans changed during mine operation, periodic deposition plans were developed, allowing more accurate closure planning for the TSF. The slurry discharge was enhanced in later years of operation by the addition of discharge pipelines and spigots around the perimeter of the TSF, as well as a floating discharge line. This allowed development of a tailings beach that would be consistent with the planned post-closure surface, which would drain to the spillway. Additional tailings were deposited in areas of the TSF with the deepest tailings to compensate for expected future settlement in those areas.

Although in a net evaporation climate, discharge of underground mine water into the TSF created a net gain of water, requiring treatment and discharge of excess water during the dry season. Water treatment plant operation also enhanced closure scheduling by reducing the volume of water in the TSF at closure and allowing the tailings beach to desiccate and gain strength during the dry season.

FINAL CLOSURE IMPLEMENTATION

A comprehensive mine site closure plan was prepared as part of initial mine permitting. Updates of the closure plan during operations allowed improvements to be made for changing site conditions and revised operating plans. The final closure plan was updated near the end of mill operations and was modified during closure based on the site conditions.

The TSF cover surface was designed and constructed as a draining surface that discharges to the spillway. The cover material was excavated from nearby borrow areas, and the minimum cover thickness was based on separation from underlying tailings, cover constructability, upward moisture migration, and root depth of preferred vegetation on the cover surface. Tailings settlement was also modeled to predict amount and rate of settlement, mainly in the deeper tailings zone. During construction, the engineer of record reviewed the cover design thickness, and it was optimized to accommodate for the material quality available in the borrow sources. The vegetation established on the cover surface is based on knowledge of preferred species in the region and establishing a full and self-sustaining plant community to provide erosion protection for the cover and discourage invasive species. As part of post-closure surveillance and maintenance, vegetation is controlled for distance of 30 m from dam slopes and from drainage channels. In these areas, vegetation is limited to grasses, to allow inspection of the cover surface.

The TSF cover has been monitored for post-closure settlement using the interferometric synthetic aperture radar (InSAR) method. Because of vegetation density on the cover, it was necessary to install 59 corner reflectors. Results have identified the zones with continued settlement (above the deepest zone of tailings) and the zones where the settlement is negligible. The InSAR monitoring has also helped to plan maintenance (soil placement and leveling) in areas of greater tailings settlement.

Because much of the upstream topography originally drained onto the TSF, temporary diversions were constructed during operation to improve the water balance. Post-closure site drainage includes diversion channels and regrading where possible to direct seasonal runoff away from or around the TSF. Remaining areas that cannot be diverted are directed onto and over the TSF cover in channels designed to convey extreme-storm runoff without eroding the TSF cover. These channels route the runoff over the cover surface and through the spillway. Drainage planning was enhanced by on-site observation during operations of erosion and sedimentation, as well as performance of diversion and drainage channels.

The performance of the TSF dams has been monitored with piezometers, survey monuments, accelerometers, and groundwater monitoring wells. Post-closure monitoring equipment has been designed and installed for long-term operation, automated data collection, and remote data review.

Closure of the mine site was aided by concurrent reclamation and pre-closure observation, which allowed observation of reclamation performance for vegetation species selection, erosion control materials, placement techniques, and sediment control. Independent technical

Courtesy of Newmont

FIGURE 2 Tailings storage facility closure conditions in March 2023, 2 years after cover completion

review was provided throughout design, construction, operation, and closure, which assisted in confirming key design parameters and improving operations and closure implementation.

Figure 2 shows the Marlin mine TSF at nearly final closure conditions. Readers are referred to Strachan and Aparicio (2020) for more details on this TSF closure. ■

REFERENCE

Strachan, C., and Aparicio, M. 2020. Case study: Pre-closure tailings management. In *Mine Tailings: Perspectives for a Changing World*. Edited by R. Furey and J. Lupo. Englewood, CO: SME. pp. 81–85.

Design and Closure Considerations, Bunkers Hill In-Pit Tailings Storage Facility, Tanami Operations, Northern Territory, Australia

Seth McCann
Newmont Australia

Mining in the Bunkers Hill open pit at Newmont's Tanami operations in the Northern Territory of Australia was completed over many campaigns between 1988 and 2003. Backfilling of the resultant pit shell with tailings commenced in 2007 in alignment with the Warlpiri traditional landowners and Central Land Council land access agreement aspirations. In 2010, a pit shell perimeter embankment was constructed to afford additional tailings containment capacity. The embankment was set back from the pit crest, blending from the natural landform elevations at the eastern and western ends with ~3 m raise in the lowest portions in between. Between 2007 and 2013, approximately 3.3 Mt (million metric tons) of tailings were deposited at a dry settled density of 1.6 t/m^3 and final footprint of ~10 ha (hectares). The general arrangement of the facility during closure design is illustrated in Figure 1.

Courtesy of Newmont Corporation

FIGURE 1 Bunkers Hill tailings storage facility general arrangement

Tailings deposition was dependant on acceptable geotechnical and hydrogeologic conditions and performance. Hydrogeologic flow modeling indicated the potential for limited localized mounding associated with increased head pressures from tailings deposition above the ground elevation and potential outward migration over time. Six pump-back wells of 70 m average depth were constructed through the perimeter embankment and operated from 2010 through the later stages of tailings deposition to mitigate the modeled aquifer influences during and post-deposition.

Previous rehabilitation success demonstrated a proven capacity to achieve safe stable landforms with established self-sustaining vegetation cover capable of faunal occupation. Hydrologic and hydrogeologic performance objectives had, however, matured since the rehabilitation of another on-site in-pit tailings storage facility (TSF), the Bullakitchie in-pit facility, in the early 2000s. When closure and reclamation planning for the Bunkers Hill TSF

Courtesy of Newmont Corporation

FIGURE 2 Bunkers Hill in-pit tailings storage facility before cover construction, October 2015

Courtesy of Newmont Corporation

FIGURE 3 Bunkers Hill in-pit tailings storage facility after cover construction, December 2016

Courtesy of Newmont Corporation

FIGURE 4 Bunkers Hill in-pit tailings storage facility after cover construction, March 2017

commenced, a design focus on achieving surface and groundwater quantity and quality completion criteria objectives was applied. The stipulated design requirements were for a landform capable of limiting groundwater recharge and mounding responses through the deposited tailings body and subsequent outward migration of poor-quality waters post-rehabilitation and decommissioning of the pump-back well infrastructure.

The completed cover design modeling, accounting for existing geochemical, climatic, physical, and hydrologic data, indicated that construction of a cover incorporating a capillary break offered the best water flux control. The selected four-layer cover system incorporated a 2-m-thick profile with a compacted layer in the middle, a non-compacted 1-m base layer constructed to promote segregation (coarser base), and a loosely placed upper layer dressed with topsoil.

Cover modeling for this modified store-and-release cover design indicated that seepage fluxes at the base of the cover approached zero such that no significant net infiltration to the tailings would occur over the long term. Modeling indicated the surface of the final cover system should be constructed to provide positive drainage with water shedding to prevent or minimize infiltration during extreme monsoonal rainfall events.

The resultant Bunkers Hill TSF final landform design was water-shedding with a water collection and diversion system (i.e., spillway) incorporated into the design. The water-shedding design was considered appropriate to mitigate risks associated with the groundwater mound, tailings geochemistry, and the native vegetation's poor tolerance of submersion. The spillway was designed with sufficient capacity to pass the critical duration probable maximum precipitation event considered to have an average recurrence interval in excess of 1 in 10,000 years and, therefore, was considered appropriate for closure design.

In 2017, the rehabilitation cover system was remodeled based on as-constructed conditions for particle size distribution, plasticity index, and phreatic conditions within the tailings body. The results of the as-built cover modeling indicate the modified 2-m store-and-release cover with a 30-cm compacted barrier over 70 cm of loose coarse waste rock would provide satisfactory long-term performance consistent with the design intent. In the 3 years post-rehabilitation, there has been a ~2 m decrease in phreatic levels within the tailings body, and performance monitoring is ongoing. Figures 2 through 4 illustrate the facility at the end of deposition (prior to cover placement), during cover placement, and post-remediation. ■

Transitioning from Mining to Solar: 21st Century Alchemy—Avalon Solar Project, Arizona, United States

Krishna Parameswaran
tfgMM Strategic Consulting

This case study advocates the siting of renewable energy projects on reclaimed tailings storage facilities (TSFs) and is based on the author's experience in coordinating the Avalon solar project, a 57-MW (megawatt) single-axis tracking solar power photovoltaic (PV) facility situated on approximately 202.3 ha (hectares) of retired agricultural land in Pima County, Arizona, near the ASARCO LLC (Asarco) Mission complex (Parameswaran 2019, 2020; Ramos and Parameswaran 2017). Asarco is strongly of the view that a solar project can be viable use of un-utilized buffer lands and post-mining use of disturbed mined lands and is committed to considering reclaimed TSFs for future renewable energy projects. Figure 1 shows an aerial view of the installation.

The Avalon solar project was born out of an Environmental Protection Agency (EPA) program, RE-Powering America's Land Initiative, which was established with the objective of siting renewable energy projects on disturbed lands, including mine sites. A unique collaboration between Asarco, the utility Tucson Electric Power (TEP), and the solar power developer Clēnera, LLC (Clēnera), made the Avalon project a reality. Clēnera was responsible for constructing the facility, which provided 300 jobs during construction. Panasonic Eco Solutions oversaw operations and maintenance of the project, and Coronal Group was the operating manager for the project. Electricity generated is sold to TEP under a 20-year power purchase agreement (PPA) helping TEP meet its clean energy goals, requiring that 70% of its power production come from renewable sources by 2035 (TEP 2023). The Avalon solar project should provide approximately $4 million in property tax revenue to Pima County over the life of the project. The solar farm commenced operation on December 23, 2014, with a Phase 1 installed capacity of 35 MW. The ribbon-cutting ceremony took place on February 27, 2015. In 2016, with implementation of Phase 2, the capacity was expanded to 57 MW.

The Avalon solar project generates enough electricity to power around 12,000 homes or roughly all the homes in the town of Sahuarita, Arizona. The generated electricity will avoid discharging roughly 93,600 tons of carbon dioxide into the atmosphere. This is equivalent to taking 19,600 cars off the road.

The Avalon project provides a road map for strategic collaboration between utilities, solar developers, utilities, and mining companies, as noted by TEP's Vice President of Energy Resources and Chief Sustainability Officer Erik Akken:

Source: Ramos and Parameswaran 2017

FIGURE 1 Aerial view of the Avalon solar project

> The Avalon systems provide TEP with enough clean energy each year to serve the annual electric needs of more than 12,000 typical homes. We're grateful the land was made available for the development of these systems, which support our ongoing transition to cleaner, less carbon intensive resources.

> Tom Murphy, mayor of Sahuarita, said:

> As the Mayor of Sahuarita, it was a distinct honor and privilege to participate in the groundbreaking and ribbon-cutting ceremonies of the Avalon Solar Project. This project was a shining example of how mining companies like ASARCO LLC can leave behind a great legacy and enhance their corporate citizenship credentials with the communities they partner with.

Mining companies have substantial land holdings, some of which are not utilized for operations. These include retired agricultural, groundwater reservoir lands acquired for water rights purposes or buffer lands. In addition, lands disturbed by mining, such as waste rock deposition areas and TSFs, are reclaimed after mining ceases. However, these reclaimed lands usually remain idle and hosting a solar facility could be a suitable post-mining land use.

The business case for considering the siting of renewable energy projects on reclaimed TSFs are as follows:

- A significant barrier in implementing renewable energy projects is the land requirement. Repurposing reclaimed TSFs and waste rock deposition facilities (WRDFs) and utilizing idle buffer lands converts a potential environmental liability into a positive legacy.
- Fostering renewable energy projects enhances the sustainability credentials of a mining company by reducing its carbon footprint in its efforts to transition to a low-carbon economy.
- Mining sites have the necessary infrastructure including transmission lines, roads, and utilities. They also have heavy earth-moving equipment that can be used for building access roads. Certain non-mineralized mined materials can be used for implementing stormwater controls.
- Mining operations represent large electricity loads and usually have medium/high voltage transmission/distribution infrastructure that can be used to transmit generated power from the solar power plant to the utility, thereby reducing interconnection costs.
- With the right regulatory framework, solar facilities developed next to mine loads can benefit from a "distributed generation" scheme, where some or even all the energy produced by the solar facility can be consumed on-site, creating efficiencies for the grid as less energy use is needed from public transmission and distribution infrastructures.

Although, to the author's knowledge, there are not yet utility-scale PV plants on reclaimed TSFs in the United States, two projects have been evaluated by the EPA under its RE-Powering America's Land Initiative for the feasibility of siting solar plants including San Xavier TSF at the Asarco Mission mine in Arizona (EPA 2011) and the Freeport-McMoRan (Freeport) Chino mine in New Mexico (Kiaatreungwattana et al. 2013).

In the United States, there are a limited number of renewable energy projects on or adjacent to mine lands, and only one, the Chevron Questa Project, is on a reclaimed TSF. Table 1 summarizes solar projects at mining company sites.

Nate Foster, the managing director of Rio Tinto Kennecott, commented that "Rio Tinto Kennecott has a key role to play in supporting the energy transition. We supply U.S. companies with the copper and tellurium they need to produce solar panels, wind turbines, and conductors. We also continue to take steps to further decarbonize our business, from our battery electric vehicle trial to our renewable diesel trial and now to our very own solar plant" (Rio Tinto 2023).

TABLE 1 Solar energy projects at mining company sites

Facility	Location (USA)	Company	Energy Type, solar	Capacity, MW	Commenced Operation
Ajo Solar Project	Ajo, AZ	Freeport	PV (photovoltaic)	5	2011
Bagdad Mine Solar	Bagdad, AZ	Freeport	PV	15	2011
Chevron Solar Project	Questa, NM	Chevron	CPV (concentrated photovoltaic)	1	2011
Avalon Solar Project	Sahuarita, AZ	Asarco	PV	35 /57	2015 and 2016
Kennecott Copper Solar Project	Magna, UT	Rio Tinto Kennecott	PV	5	2023

Source: Ramos and Parameswaran 2017; Rio Tinto 2023

U.S. copper companies are committed to transitioning to a low-carbon economy. Asarco is committed to achieving 15% Scope 1 and 2 greenhouse gas emissions by 2027 (2018 baseline emissions); 35% reduction by 2035; and achieving net zero emissions by 2050 (Grupo México 2022). This is to be accomplished by transitioning to electric-powered mine trucks and investments in renewable electricity and nature-based solutions, including carbon sequestration by reforestation and wetlands creation.

Freeport is evaluating a large utility-scale installation at one of its Arizona mine locations as part of its Clear Skies initiative whose objective is to integrate up to 450 MW in renewable power projects (wind/solar/battery storage) and PPAs in the United States (Freeport-McMoRan 2022). The project is in the planning stages with decision on the project several years off.

The two challenges facing the mining industry are intense global scrutiny from stakeholders as a result of continuing catastrophic failures of TSFs and the need to transition to a low-carbon economy to combat climate change. Coupled with safer management of existing and legacy TSFs as a result of implementation of the Global Industry Standard on Tailings Management (GISTM) (GTR 2020), these two formidable challenges can be converted into a great opportunity by repurposing reclaimed TSFs and WRDFs. It is the author's fervent hope that siting renewable energy plants on reclaimed TSFs and WRDFs can help transform the U.S. Southwest and México's Sonora region into the solar capital of the world and showcase the contributions of mining companies toward combating climate change. It is the way for mining companies to be part of the solution rather than part of the problem. ■

REFERENCES

EPA (U.S. Environmental Protection Agency). 2011. *Renewable Energy Development Opportunities: ASARCO Mission Mine Tailings Area, San Xavier District, Tohono O'odham Nation.* Washington, DC: EPA.

Freeport-McMoRan. 2022. *Climate Report: The Power of Copper.* pp. 26–27. https://view.publitas.com/freeport-mcmoran/2022-climate-report. Accessed November 2023.

Grupo México. 2022. Sustainable Development Report. pp. 215–263, 421–422. https://www.gmexico.com/GMDocs/InformeSustentable/Eng/SDR_2022.pdf. Accessed November 2023.

GTR (Global Tailings Review). 2020. *Global Industry Standard on Tailings Management.* https://globaltailingsreview.org/wp-content/uploads/2020/08/global-industry-standard_EN.pdf. Accessed November 2023.

Kiaatreungwattana, K., Geiger, J., Healey, V., et al. 2013. *Feasibility Study of Economics and Performance of Solar Photovoltaics at the Chino Mine in Silver City, New Mexico.* NREL/TP-7A30-57959. Golden, CO: National Renewable Energy Laboratory.

Parameswaran, K. 2019. Sustainable development and the primary copper industry—Past, present and future. In *Proceedings of the 58th Conference of Metallurgists (COM 2019) Hosting Copper 2019.* Westmount, QC: Canadian Institute of Mining, Metallurgy and Petroleum. pp. 1–20.

Parameswaran, K. 2020. Case study 15: Innovative post-mining land use: Siting renewable energy projects on reclaimed tailings storage facilities. In *Tailings Management Handbook: A Life-Cycle Approach.* Edited by K.F. Morrison. Englewood, CO: SME. pp. 918–919.

Rio Tinto. 2023. Rio Tinto Kennecott completes construction of solar power plant. News release. October 10. https://www.riotinto.com/en/news/releases/2023/rio-tinto-kennecott-completes-construction-of-solar-power-plant. Accessed November 2023.

Ramos, M., and Parameswaran, K. 2017. 21st century alchemy: Transitioning from mining to solar. In *2017 Sustainable Industrial Processing Summit.* Volume 1: Barrios International Symposium. Edited by F. Kongoli, M. Palacios, T. Buegner, et al. Mont-Royal, Quebec, Canada: Flogen Star Outreach. pp. 161–177.

TEP (Tucson Electric Power). 2023. A cleaner, greener grid. https://www.tep.com/renewable-resources-2/. Accessed November 2023.

Open Pit Co-Deposition Closure Backfill, Marlin Mine, San Marcos, Guatemala

Manuel Aparicio
Newmont Corporation

Marlin mine is located approximately 120 km west-northwest of Guatemala City, in a mountainous region of western Guatemala. Mining and milling operations at the mine commenced in late 2005, including open pit and underground mining. Open pit mining was phased out in 2012, while underground mining continued to the end of milling operations in mid-2017. Processing for precious-metal recovery utilized conventional milling, with a cyanide leach process and Merrill–Crowe circuit for the recovery of gold and silver. Mill production averaged 4,500 t (metric tons) per day.

The area is characterized by high seismicity, and the regional climate is tropical, reflecting distinct wet and dry seasons with an average annual precipitation and evaporation of approximately 1,200 and 1,700 mm, respectively. Average maximum and minimum temperatures are 25°C and 10°C, with an average annual temperature of 18°C.

PIT BACKFILL CLOSURE SOLUTION

Mill tailings were initially deposited into a conventional tailings storage facility (TSF) in a valley downstream of the mill, designed and operated to store mill tailings slurry and process water (see the case study, "Closure of a Conventional Tailings Storage Facility in a Seismically Active Tropical Environment, Marlin Mine, San Marcos, Guatemala"). However, the ultimate TSF capacity did not meet life-of-mine tailings production volumes. The TSF had already been expanded once, and building a new TSF was not considered a viable option. Marlin's team developed a solution to filter and stack tailings as pit backfill.

A comprehensive mine site closure plan was prepared as part of the initial mine permitting. The closure plan was updated several times during mine operation as site conditions changed. As such, this scenario presented an opportunity to update the closure plan for the open pit so that filtered tailings could be conveyed and/or trucked to the open pit for partial backfill. The updated closure plan for the open pit was issued and approved by Guatemalan regulators. The filter plant, using pressure filters, was commissioned in 2012, just as the Marlin open pit was finishing operations.

The tailings management plan was also modified as filtered tailings were used for backfill in the open pit in combination with underground mine waste rock, with the TSF as a secondary deposition location for tailings slurry. The tailings deposition plan was evaluated monthly, and a percentage split was established depending on the mine plan tonnages, filter and process plant availability, and remaining TSF capacity.

Courtesy of Newmont

FIGURE 1 Marlin pit at the beginning of backfill operations, January 2013

A total of 1.85 Mm^3 (million cubic meters) of filtered tailings were placed to backfill the pit, together with 3 Mm^3 of waste rock and 0.9 Mm^3 of soil cover. This backfill (up to 89 m thick) produced a terrace level and a 60-m high buttress for the pit highwalls, improving the stability of the pit highwall from a fault that exhibited considerable displacement during mining (Figure 1).

OPERATIONAL CHALLENGES

The general design configuration for the pit backfill comprised a 10-m-wide rim of filtered tailings against the pit walls with waste rock placed inside the filtered tailings rim. The materials were placed in lifts with thicknesses of approximately 30 cm for filtered tailings and 90 cm for waste rock. The filtered tailings backfill required sufficient bearing capacity for trafficability conditions during placement, especially during rainy seasons. Because the bearing capacity of the filtered tailings decreased with increasing moisture content, and the clay-rich condition of the tailings caused it to hold more moisture, a solution for wetter than optimum filtered tailings was needed. Approximately 70% of the tailings gradation was fine-grained (passing 0.075 mm) (Figure 2), and the average (2012–2017) filtered tailings liquid limits and plastic limits were 33 and 23, respectively, with a plasticity index of 10. A series of laboratory and field tests were developed to find the mixture of filtered tailings amended with cement that would maintain a 19% moisture content, compared to an average 14% (range from 12%–16%) optimum moisture content obtained from standard Proctor tests, with an undrained shear strength of 100 kPa required for acceptable bearing capacity.

The test results showed that a mixture of 3% cement by weight would achieve the desired strength and moisture content. Within the first 6 months of full-scale operation, field testing demonstrated that the filtered tailings did not require cement addition during the dry season and that only 1% cement by weight was sufficient to achieve the parameters needed for trafficability during the rainy season. Historic data show that during the high-precipitation months of the rainy season and when the filter plant presented upset conditions, up to 1.5% cement was added. An average of 0.5% cement was used during lower precipitation months. Throughout

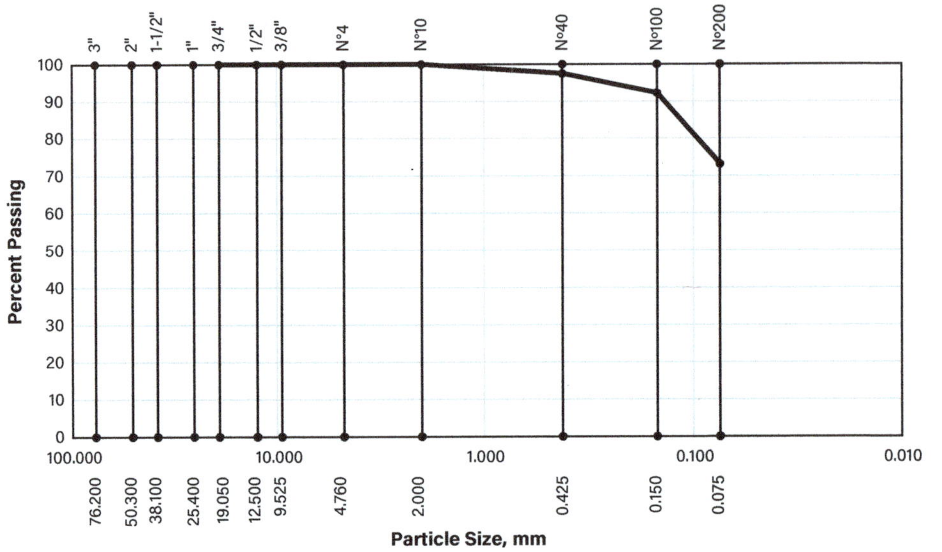

Courtesy of Newmont

FIGURE 2 Typical grain size distribution for filtered tailings

TABLE 1 Achieved tailings in-place*

	Tailings In-Place Properties			
	Percent Compaction	In-Place Density, t/m³	Percent Moisture Content	Shear Strength, kPa
Specification	80.0	1.75 ± 0.05	20.0	100.0
Average (2012–2017)	86.1	1.7	19.5	119.1

* Results shown are averaged from quality control data.
Courtesy of Newmont

operations, approximately 0.1% lime was also added to the filtered tailings. Table 1 summarizes the achieved tailings in-place characteristics as compared to the specifications.

A temporary storage building was included in the design of the filter plant, which allowed the mixture of cement and filtered tailings to cure for at least 12 hours before placement in the pit. The tailings moisture content was checked at the filter plant, within the storage building, and after placement and compaction in the pit. This information aided the operators in adjusting the filtering times during the dry season and the cement content during the rainy season.

Other challenges included:

- Surface water management during fill placement
- Water sprays installed to control dust at the filter plant
- Alteration of filter cloth material from the original design to allow a more efficient filtering process because of the high clay content of the tailings

BENEFITS OF THIS CLOSURE SOLUTION

Although there were operational challenges, the co-deposition of filtered tailings with waste rock as pit backfill provided a unique solution to address the closure of the open pit at the

(A) November 2019

(B) August 2023

Courtesy of Newmont

FIGURE 3 Marlin pit final closure conditions in (A) 2019 and (B) 2023

Marlin mine. Specifically, this closure approach allowed development of a free-draining cover and partial backfill of the open pit to encapsulate mineralized materials on the pit walls and improve slope stability. As illustrated in Figure 3A and 3B, this unique solution facilitated prompt closure of the facility in a tropical climate despite a combination of high clay-content tailings used as pit backfill and the potential for acid generation from exposed pit walls and mine waste rock.

After completing closure of the facility, a monitoring method was implemented using interferometric synthetic aperture radar (InSAR), which allows the rate of movement that may occur over time to be recorded. The data has shown that the structure maintains its stability, has not presented settlement that allows ponding of runoff water on the surface and slopes, the drainage systems remain functional, and no maintenance of any kind has been necessary. ■

Repurposing Tailings and Other Earth Material Waste into Ceramics

Ross Guenther
Ceramext, LLC

Over the past 20 years, innovative technologies have been developed to repurpose tailings and mine waste into many products. Developers, working with scientists and engineers, have created the technology to produce high-quality, low-cost ceramic tile and other building products from raw materials including mine tailings, quarry fines, biomass ash, coal fly ash, and other waste materials. Most mining companies are only interested in mining and not in producing ceramics, although this may be changing. Often, mineral prospects are in areas without suitable topography for tailings facilities or in areas of high land costs. And many mines require fine grinding (to –100 mesh) to recover an economic product while paying for the costs of environmentally safe tailings management.

HOT VACUUM EXTRUSION OF CERAMICS

Although ceramic engineers were taught it was impossible, while studying the progression of heating tailings in 5°C increments, Ceramext engineers noticed that because certain minerals melted first, if compressed at an appropriate temperature, then a coherent, strong, impermeable ceramic could be produced. Analyzing hundreds of thin sections with a scanning electron microprobe, the engineers found that fractures within the grains of the tailings extended the supporting glass from the outer glassy matrix into the mineral grains, which greatly strengthens the ceramic while still allowing it to be pressed and shaped, or even cooled and later reformed. Some of the minerals to create more favorable stronger glassy components include Na and K feldspars and mica found in most copper mine tailings and some gold mine tailings.

The first 10,000 hours on this project were spent building and patenting an apparatus for hot vacuum extrusion of ceramics. Tailings were heated within a kiln at various temperatures, generally within a 5-cm pipe, then pushed out with a 10-ton press. Later, for most purposes, the vacuum was not required. Tailings and fly ash from more than 50 locations were tested.

Ceramext continued operations in 2003 at the Idaho-Maryland mine site in Grass Valley, California, historically one of the largest gold mines in the United States. Using the gold tailings, other sources of tailings, and the aggregate fines from Granite Rock Company's quarry, which is the largest aggregate quarry in California, several tons of high-quality tile were produced.

Successful testing was completed for producing tile from several Arizona copper mine tailings. And working with the Pacific Northwest National Laboratory, a partially melted basalt from local basalt quarry waste fines was used to produce building materials that can be extruded into large rectangular and cylindrical shapes. This method is particularly applicable for extruding cylindrical log-shape ceramics to encapsulate radioactive waste. In addition, the

National Aeronautics and Space Administration (NASA) has been studying the use of building materials using regolith on the moon and Mars.

The hot forging process was refined: heating and partially melting tailings materials to approximately 1,000°C, applying 6,900–13,800 kPa compression, and cooling. The strength of the material can be increased by using pressure while the material is hot, but at temperatures lower than the complete melting point. This process is shown in Figure 1.

Generally, a cold steel die is used, and as the die slightly expands from the heat of the ceramic, it facilitates the ejection of the hot-forged tile. With minor modifications, conventional green ceramic presses and electric or gas roller kilns can be used for hot forging. Porcelain-quality tiles can be made by only screening raw materials without the expensive grinding and spray-drying steps. No binders are required, and up to 100% of the waste material can be used. New advanced ceramic compositions using this hot forging technology have been patented.

By repurposing 100% mine tailings from gold mines, copper mines, aggregate quarries, and other waste materials, ceramics have been produced in several successful pilot plants. These pilot plants have produced tile as well as other ceramic products including floor tile, roof tile, wall cladding, pavers, bricks, and countertops. As shown in Figure 2, the tiles can also have the appearance of a wide variety of natural rock surfaces and can be polished or glazed if desired. Approximately 50 different types of recycled feed materials have been tested. Floor, wall, and roofing tile, and porcelain-grade countertops have been produced with a range of attractive earth-tone colors. The tiles shown in Figure 2 were made with a

Courtesy of Ceramext, LLC

FIGURE 1 Hot forging technology

Courtesy of Ceramext, LLC

FIGURE 2 Tile finishes made from mine tailings and coal fly ash

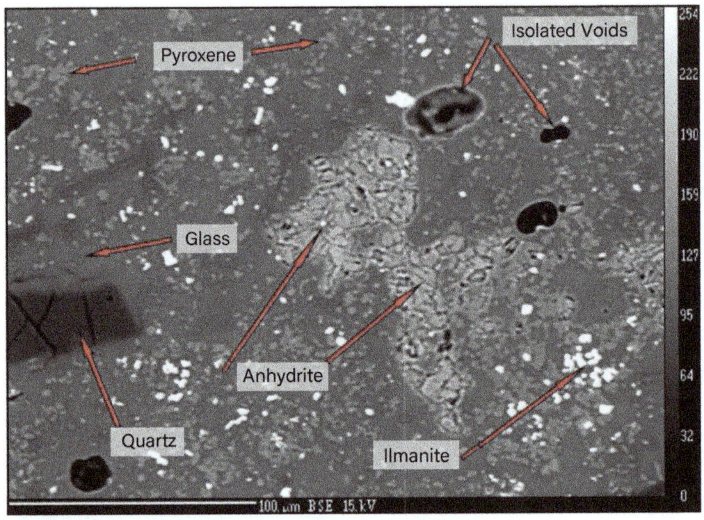

Courtesy of Ceramext, LLC

FIGURE 3 Tile thin section from Idaho-Maryland mine shown via scanning electron microprobe

strong impermeable porcelain quality, with no additional additives. Tile colors possible without additives include white, black, beige, and subtle greens. Mineral additives can also produce many combinations of natural stone appearances and textures. Glazes in multiple colors can be applied, if desired. Relief designs can also be made.

Composition patents consist of partly melted rock fragments (clasts), glass melted from the clasts, and crystallites formed in the new glass. Tests on tiles from the Idaho-Maryland mine have a modulus of rupture of 69,000 kPa, much higher than typical Italian porcelain. The tiles are impermeable to water, even without glazing. These tiles are scratch resistant with a Mohs hardness of about 8, harder than quartz. The unmelted clasts act much like the aggregate in concrete with the newly formed glass as the cement binder holding the clasts together and with the crystallites further reinforcing the glass itself. The contaminants, if any, are generally rendered inert to leaching or other means by locking them within a stable glassy matrix in the tile (see the thin section in Figure 3). The initial composition patent had to also include anhydrite ($CaSO_4$), which makes the sulfur stable as water cannot reach it, creating gypsum. Similar, if not the same, anhydrite is now used on the remote sensors (for sulfur) on the NASA Mars rovers. ■

PART IV

Environmental and Regulatory

Evolution of Brazilian Tailings Regulations in Response to Failures

Kimberly Finke Morrison
Newmont Corporation

Brazil is the world's third largest country with respect to geographical area and possesses rich reserves of important commodities. Ranking as the third largest global producer of both iron ore and bauxite (Brown et al. 2018), and the largest global producer of niobium, far exceeding any other country in the world (Ronald 2016), Brazil clearly ranks among the largest nonfuel mineral producers in the world today. As of 2018, Brazil had an estimated 839 tailings storage facilities (TSFs) with approximately 44% located in the State of Minas Gerais (ANM 2018; Morrison et al. 2018). Given the global importance of mining to the country, it is not surprising that Brazil has experienced several notable tailings embankment failures (Morrison et al. 2018). What is surprising, however, is that regulation pertaining to tailings embankment safety is a relatively new concept in the country.

TAILINGS STORAGE FACILTY FAILURE HISTORY IN BRAZIL

Up to and including the failure of Dam I at the Córrego do Feijão iron ore mine near Brumadinho, Brazil, on January 25, 2019, which resulted in 270 fatalities, there have been at least nine notable tailings embankment failures documented in Brazil since 1986 (Morrison and Gomide 2019). Many of the failures had severe consequences, and most occurred before the advent of Brazilian legislation applicable to tailings embankments (Morrison et al. 2018). Figure 1 presents a timeline of major tailings embankment failures in Brazil and implementation of key TSF safety regulation.

The first documented tailings embankment failure in Brazil occurred in 1986 at the Fernandinho iron ore mine, resulting in seven fatalities (Barbosa 2015). No other documented TSF failures occurred until 2001 when a tailings embankment owned by Mineração Rio Verde ruptured in Macacos, a district in the municipality of Nova Lima in the metropolitan area of Belo Horizonte (Minas Gerais), resulting in five fatalities and significant environmental damage (Barbosa 2015). In 2007, the Rio Pomba bauxite mine (Minas Gerais) was shut down after heavy rains caused the tailings embankment to break, releasing approximately 2 million m^3 of red mud, partially burying streets and houses in two neighboring villages, and leaving about 4,000 residents homeless (Morrison et al. 2018). Remarkably, this failure resulted in no fatalities, which was believed to be the result of an effective emergency response plan (Morrison et al. 2018). The 2007 tailings embankment failure at Rio Pomba followed a leakage event that had taken place at the same facility in March 2006, and though the public was informed that the mud was nontoxic, the mining company was forced to shut down (Reuters Staff 2007). On November 5, 2015, the Fundão tailings embankment at the Samarco iron ore mine (Minas

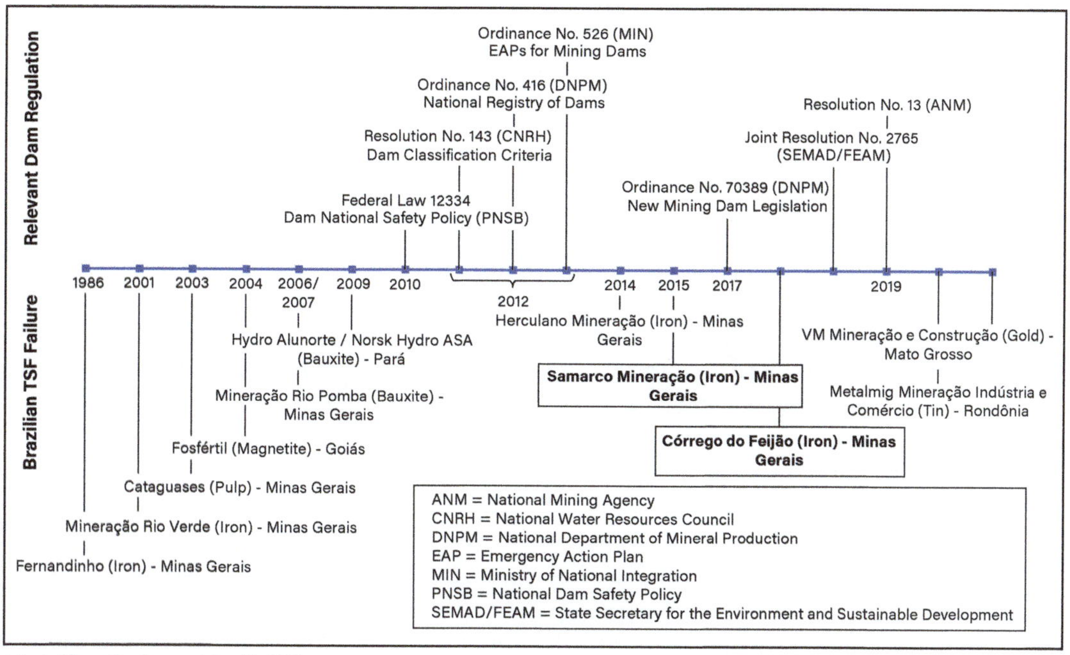

Adapted from Morrison and Gomide 2019

FIGURE 1 Chronological sequence of reported tailings storage facility (TSF) failures in Brazil and notable changes in regulation

Gerais) suddenly failed, releasing an estimated 43 million m^3 of tailings (80% of the total contained volume), polluting 668 km of watercourses extending from the Doce River to the Atlantic Ocean, and killing 19 people (Carmo et al. 2017).

FIRST NATIONAL DAM SAFETY POLICY

The first dam safety policy was published in September 2010 via Brazilian federal Law No. 12,334 (Presidency of the Republic 2010). This law established the National Dam Safety Policy (PNSB), which applies to any dam intended to accumulate water and/or waste for any use meeting one or more of the following four conditions:

1. Dam height greater than or equal 15 m
2. Reservoir capacity greater than or equal to 3 million m^3
3. Reservoir containing hazardous solids (according to Brazilian standards)
4. Classification of medium or high with respect to associated potential damage

Law No. 12,334 set out general requirements for dam owners with respect to dam safety, including classifying embankments based on risk, development of dam safety plans, and dam safety inspections and reviews. As the law was general in nature, subsequent regulation was developed to provide additional detail:

- Resolution No. 143 was published by National Water Resources Council (CNRH, Conselho Nacional de Recursos Hídricos) in July 2012 to establish criteria for dam classification based on risk category and potential associated damage (CNRH 2012).

- In September 2012, the National Mining Agency (ANM, Agência Nacional de Mineração), formerly the National Department of Mineral Production (DNPM, Departmento Nacional de Produção Mineral [DNPM 2012]), published Ordinance No. 416 that created the National Mining Dams Registry and provided requirements for periodic safety reviews and both regular and special safety inspections.
- In September 2012, Ordinance No. 526 was published (MIN 2012), which specifically addressed emergency action planning for mining dams.

REGULATORY CHANGES AFTER SAMARCO

The failure of Samarco's Fundão tailings embankment in November 2015 effected significant changes to the regulations and requirements surrounding tailings management in Brazil. ANM published Ordinance No. 70,389 in May 2017 (DNPM 2017), which modified the National Mining Dams Registry; revised the classification criteria for tailings embankments; required real-time monitoring of installed dam instrumentation, including continuous video monitoring; overhauled the emergency action plan requirements; established requirements for periodic dam safety reviews; and put forth other requirements (Morrison et al. 2018). Operators had until May 2019 to fully comply with the new law. This ordinance also required deployment of sirens and other appropriate mechanisms for warning the population in the self-rescue zone (ZAS) downstream of the TSF of imminent failure, whereby the *ZAS* is defined as the greater of (1) the region downstream of the embankment within 10 km; or (2) where the time of arrival of a flood wave based on a hypothetical dam breach analysis is equal to 30 minutes (Morrison et al. 2018).

The dramatic video footage of the failure at Brumadinho may be attributed to Ordinance No. 70,389, which required installation of continuous (24-7) video monitoring of tailings embankments. Although sirens were installed in the area downstream of Dam I in accordance with this ordinance, the system was either not functioning or not triggered to warn people downstream.

CONTINUED EVOLUTION OF REGULATIONS AFTER BRUMADINHO

Mere days after the failure at Brumadinho, new regulations were imposed in the State of Minas Gerais, while a couple of weeks later, more strict regulations were proposed for tailings management within the entirety of Brazil (Morrison and Gomide 2019). On January 30, 2019, Joint Resolution SEMAD/FEAM [State Secretary for the Environment and Sustainable Development] No. 2,765 was published in the Minas Gerais *Official Gazette* (Antunes and Gonçalves 2019). The joint resolution required de-characterization of all tailings embankments constructed using the upstream method or an unknown method in the State of Minas Gerais, whereby *de-characterization* is defined as the process in which the embankment no longer has the characteristics of a dam (i.e., does not operate as a tailings containment structure) and is destined for another purpose (Antunes and Gonçalves 2019).

In mid-February 2019, ANM (2019a) published draft Resolution No. 4, which was finalized in August 2019 via Resolution No. 13 (ANM 2019b). Resolution No. 13 banned future use of the upstream construction method and required de-characterization of all existing upstream-constructed tailings embankments. De-characterization includes decommissioning, physical and chemical stabilization, hydrologic and hydrogeologic control, and monitoring for the period necessary to confirm effectiveness (ANM 2019b). Additionally, Resolution No. 13

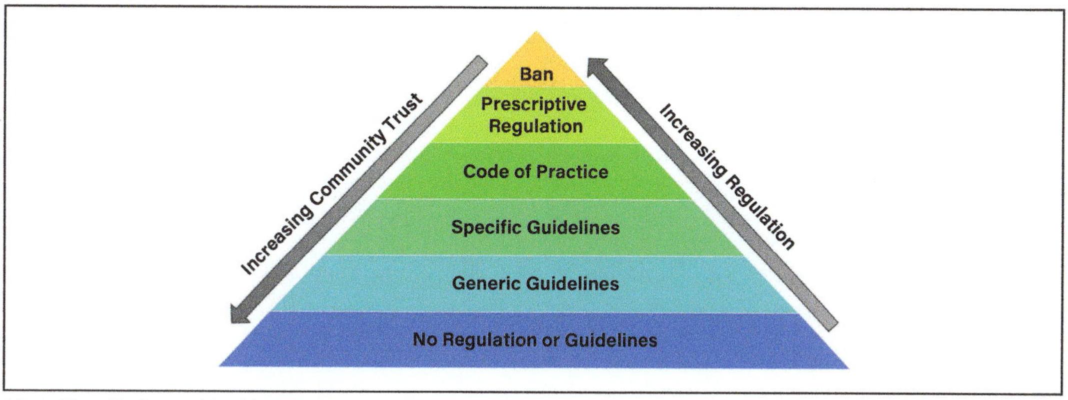

Adapted from Morrison and Gomide 2019

FIGURE 2 Trust-control relationship

provided other requirements regardless of embankment construction methodology, including (ANM 2019b):

- Prohibits maintaining buildings within the self-rescue zone (10 km downstream of the TSF)
- Prohibits construction of other dams or storage of radioactive waste in the downstream area
- Requires automated instrumentation with full-time monitoring for TSF embankments with an associated potential damage characterization of high
- Establishes minimum factor of safety requirements
- Requires water reduction measures

TRUST-CONTROL RELATIONSHIP

Failures and incidents are often the catalyst for change, and this is evident with tailings embankment failures and the inherent increase in regulatory control (Morrison 2020, 2021) as experienced in Brazil since 2010 (Morrison et al. 2018; Morrison and Gomide 2019). Failures degrade public trust, resulting in a corresponding increase in regulatory controls as demonstrated by the trust–control relationship illustrated in Figure 2. ■

REFERENCES

ANM (Agência Nacional de Mineração [National Mining Agency]). 2018. Classification of Mining Dams Database December 2016. www.anm.gov.br/assuntos/barragens/plano-de-seguranca-de-barragens. Accessed June 2018.

ANM (Agência Nacional de Mineração [National Mining Agency]). 2019a. Resolution No. 4, February 15. *Official Gazette of the Federal Union*, Brazil.

ANM (Agência Nacional de Mineração [National Mining Agency]). 2019b. Resolution No. 13, August 8. *Official Gazette of the Federal Union*, Brazil.

Antunes, P.d.B., and Gonçalves, V. 2019. Brazil: Joint Resolution SEMAD/FEAM No. 2.765/2019: De-characterization of upstream tailings dams in Minas Gerais. Mayer Brown (February 27). www.mayerbrown.com/en/perspectives-events/publications/2019/02/joint. Accessed February 2021.

Barbosa, F. 2015. Worst dam accidents in Brazil's Minas Gerais. *Business News Americas*. November 13. www.bnamericas.com. Accessed February 2021.

Brown, T.J., Idoine, N.E., Raycraft, E.R., et al. 2018. *World Mineral Production, 2012–2016*. Keyworth, Nottingham, UK: British Geological Survey.

Carmo, F.F., Kamino, L.H.Y., Junior, R.T., et al. 2017. Fundão tailings dam failures: The environment tragedy of the largest technological disaster of Brazilian mining in global context. *Perspectives in Ecology and Conservation* 15:145–151.

CNRH (Conselho Nacional de Recursos Hídricos [National Water Resources Council]). 2012. Resolution No. 143, July 10.

DNPM (Departmento Nacional de Produção Mineral [National Department of Mineral Production]). 2012. Ordinance No 416, September 3. Revoked and replaced with Ordinance No 70.389.

DNPM (Departmento Nacional de Produção Mineral [National Department of Mineral Production]). 2017. Ordinance No 70.389, May 17.

MIN (Ministério da Integração Nacional [Ministry of National Integration]). 2012. Ordinance No 526, September 6.

Morrison, K.F. 2020. Tailings dam failures and regulatory or social response: A reflection of lack of trust? In *Mine Tailings—Perspectives for a Changing World.* Edited by R. Furey and J.F. Lupo. Englewood, CO: SME. pp. 95–102.

Morrison, K.F. 2021. A reflection of lack of trust? Tailings dam failures and regulatory or social response. *Mining Engineering* 73 (January): 20–25.

Morrison, K.F., and Gomide, P.C.R. 2019. Post-Córrego do Feijão: Continued evolution of tailings dam regulations in Brazil. In *Tailings and Mine Waste 2019: Proceedings of the International Conference on Tailings and Mine Waste.* Edited by J. Goodwill, D. van Zyl, and M. Davies. Vancouver, Canada: University of British Columbia. pp. 391–400.

Morrison, K.F., Pedrosa, H.G., dos Santos, G.J.I., et al. 2018. Changes to tailings dam regulations in Brazil in the aftermath of failures. In *Tailings and Mine Waste 2018: Proceedings of the International Conference on Tailings and Mine Waste.* Vancouver, Canada: University of British Columbia. pp. 27–34.

Presidency of the Republic. 2010. [Brazil] Law No. 12.334, September 10. [In Portuguese.]

Reuters Staff. 2007. Brazil bauxite mine closed after dam burst. January 21. www.reuters.com/article/idUSN10315283. Accessed February 2021.

Ronald, E. 2016. Niobium in a nutshell. *Mining Geology HQ.* May 28. www.mininggeologygroup.com/niobium-in-a-nutshell/. Accessed February 2021.

Surface Deposition of High-Sulfide Paste Tailings, Neves Corvo Mine, Portugal

Rens Verburg
WSP USA

Neves Corvo is an underground copper-zinc mine located in the south of Portugal. The mine has been operated since 1988 by Somincor. The tailings produced by the operation contain approximately 40%–50% pyrite and, as such, have a very high acid generation potential. Since mine start-up, they have been subaqueously placed in a tailings storage facility (TSF). The subaqueous disposal system has provided safe and reliable storage of tailings. However, because of the finite capacity of the existing TSF, alternatives for tailings placement were studied by Somincor. Surface deposition of paste was identified as the preferred option, and paste is currently placed in individual cells within the existing TSF, enabling progressive filling and concurrent reclamation of the facility.

The feasibility of changing the tailings deposition method to subaerial placement of paste has been investigated since 2001. Work completed included a geochemical characterization program of the paste tailings, a laboratory screening program, a 3-year field cell monitoring program, a large-scale field trial, a feasibility study and impact assessment, basic engineering, detailed engineering, commissioning of the paste plant, and construction of the first four operational paste cells.

Understanding the geochemical behavior of the sulfide-rich tailings was paramount as subaerial deposition methods can promote significant generation of acid rock drainage. Therefore, three phases of testing were undertaken. The first phase was conducted in the Somincor laboratory and involved a screening-level assessment of different subaerial placement options for many materials with different moisture contents and amendments (cement, lime, bactericide).

The study demonstrated the beneficial effects of higher moisture contents with respect to sulfide oxidation. The wettest samples consistently outperformed the drier samples by exhibiting the longest lag times to acid generation. The observed behavior of amended samples was also consistent with expectations. The lime and cement provided early buffering capacity, but in the longer term, little difference was observed between the amended and unamended samples because the neutralizing capacity of these amendments was not sufficient to offset the acid-generating capacity of the sulfides.

The study expanded to the field in 2002 with the construction of six field cells, each containing approximately 12 t (metric tons) of tailings paste with and without amendments. The cells were monitored for 3 years, with routine collection and comprehensive chemical analysis of seepage and runoff.

The results from this program confirmed sulfide oxidation occurs in the paste. However, because of its low permeability, contaminant transport was very slow, and the impact of sulfide oxidation on seepage quality was reduced through the presence of mineralogical controls that buffered pH. It was further identified that, during active paste deposition, it would be the control of runoff, not seepage, that would govern the placement protocol. A lag time to acid generation was established that represented the operational window within which a previously placed layer of paste needed to be covered by a fresh paste layer or a cover to prevent an unacceptable degree of sulfide oxidation.

In 2005, a paste plant was commissioned in support of construction of a 50,000 m^3 pilot paste trial. This trial was initiated to investigate paste deposition techniques, co-deposition with acid-generating waste rock, and long-term performance of four configurations (Figure 1):

- **Bentonite-enriched sand (BES).** A barrier cover consisting of an engineered subsoil (clean quarry rock) underlain by potentially acid-generating (PAG) waste rock, separated by a BES layer

FIGURE 1 Paste trial

Google Earth

FIGURE 2 Aerial overview of operational paste cells

- **No capillary break.** A low-flux cover consisting of an engineered subsoil underlain by PAG waste rock
- **Capillary break.** A low-flux cover consisting of an engineered subsoil underlain by PAG waste rock, with the subsoil and waste rock separated by a capillary break layer
- **No cover.** Control cell

Monitoring points were installed into the paste and covers for collection of water samples from the unsaturated and saturated zones. Data interpretation involved extensive geochemical modeling, unsaturated flow modeling, and an impact assessment to evaluate seepage quality and quantity and potential associated effects on the receiving groundwater aquifer. Based on the findings of this program, a low-flux cover was identified as the preferred alternative to maintain a certain degree of saturation of the underlying paste.

The operational paste deposition method required the construction of a paste plant. Paste is generated at a rate between 150 and 250 t/h with a target solids content between 62% and 72%. The partitioning of the existing TSF in cells is achieved by constructing internal rock-fill dikes (consisting of PAG waste rock) on top of the existing slurry tailings. Lime is added to the facility to maintain neutral conditions in any surface water during operation. When the cell filling is complete, the low-flux cover layer will be applied on the paste to assure its chemical and physical stability. This methodology enables progressive filling and concurrent closure of the facility and allows for maintaining the existing perimeter embankment at its current elevation. Figure 2 shows an aerial overview of recent conditions. ■

Systematic and Sustainable Tailings Storage Facility Dust Control, Peñasquito Mine, Mexico

Ross N. Hunsaker
Newmont Corporation

Dust control is a universal challenge for tailings storages facilities (TSFs) worldwide. At the Peñasquito mine site in Zacatecas, Mexico, the challenge is intensified because the TSF is more than 11 km in perimeter, resides in a semi-arid desert climate, and uses the coarse portion of the tailings (i.e., cyclone underflow) as construction material. Even with these challenges, Peñasquito has demonstrated considerable improvement by implementing a systematic and sustainable approach to dust control in its operation at the TSF, and this case study explains how the team achieved these results, what the associated costs are, and what drove the decision to improve.

The Peñasquito TSF is a centerline raise embankment that uses the coarse portion of the tailings stream to build the structural portion of the embankment in a paddock-style construction, using waste rock from the open pit as a rock shell. For the construction activities on the crown, sand is hydraulically placed in 2.5-m-deep paddocks. The primary sources of dust at the Peñasquito TSF are haul roads, quality sand section, and the tailings beach. Two relatively simple strategies used at Peñasquito are to use the spigoting plan to maintain humid beaches and route the traffic on the crown to avoid disturbing the quality sand section. A significant effort that resulted in a sizable step change for reduction in dust generation at Peñasquito occurred when dust control was incorporated as a part of operations' daily activities (Figure 1). After dust suppression is applied, the area is marked as an exclusion zone with the use of stakes and signage to reduce the likelihood of any disturbance of the crust formed through the application of the dust suppressant.

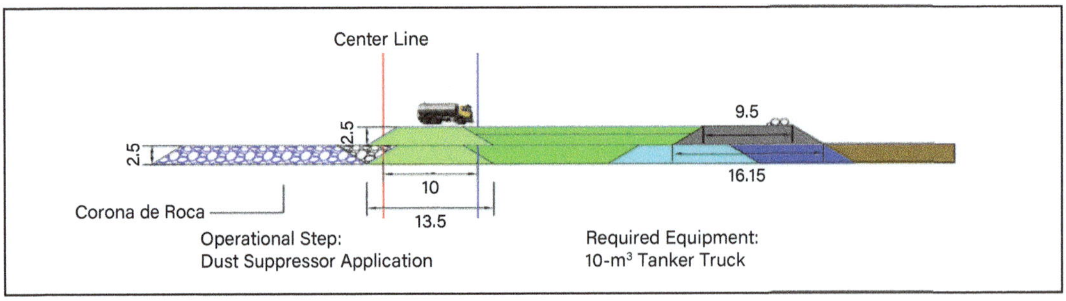

Courtesy of Newmont

FIGURE 1 Step 6 in the crown construction sequence is the application of dust suppression

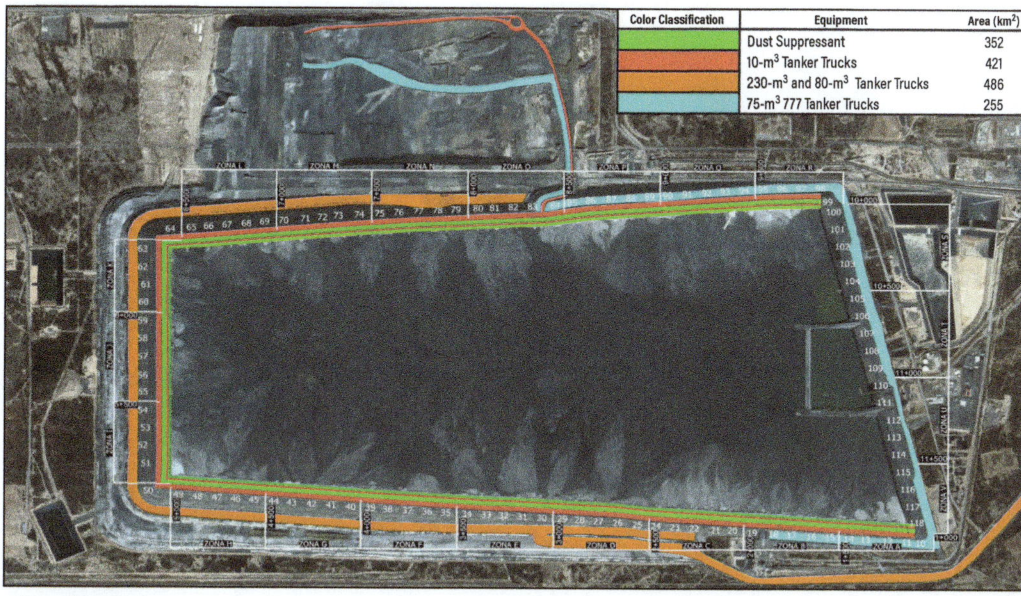

Courtesy of Newmont

FIGURE 2 Location of dust control, equipment requirements, and surface area approximation at the Peñasquito tailings storage facility

Figure 2 shows an aerial view of the Peñasquito TSF with the location of dust generation and controls, equipment requirements, and surface area approximations. The orange line represents the 300-t (metric ton) truck haul road. Dust on this road is managed using water applied from 230-m^3 and 80-m^3 water trucks. The teal line represents the 100-t truck haul road. Dust on this road is managed using water applied from a CAT 785 water truck with a capacity of 75 m^3. The red line represents the rock fill on the crest. Dust from this area is managed using water applied from a 10-m^3 tanker truck. The green line represents the quality sand section, and dust generated from this area is controlled by applying a polymer-based dust suppressant using a 10-m^3 tanker truck.

The implementation of the control measures previously described has resulted in a consistent and sustainable dust control system for more than 3 years as can be seen in Figure 3. The figure also shows that the average monthly concentration of PM_{10} (particulate matter 10 µm or less in diameter) was below the Mexican standard of 75 mg/m^3. This is a significant accomplishment for the TSF team at Peñasquito, given that in previous years, there were multiple community complaints related to excess dust. In 2019 and 2020, there was only one documented complaint per year from the nearby communities with respect to dust.

The cost for dust control measures on the tailings crown is a small fraction of the construction cost. Costs for dust control at the Peñasquito crown fill include the purchase of the dust suppressant and the operation of the suppressant and water trucks (10-m^3 tankers). It is critical that these activities be included as a line item in the annual budgeting process. In Peñasquito's case, dust control represents approximately 4% of the total cost of constructing and operating a TSF. While it is difficult to put a direct economic value on improved community relations, if a credible and trustworthy relationship exists between the operator and the community, then it certainly is plausible to state that future permits, expansions, or community negotiations

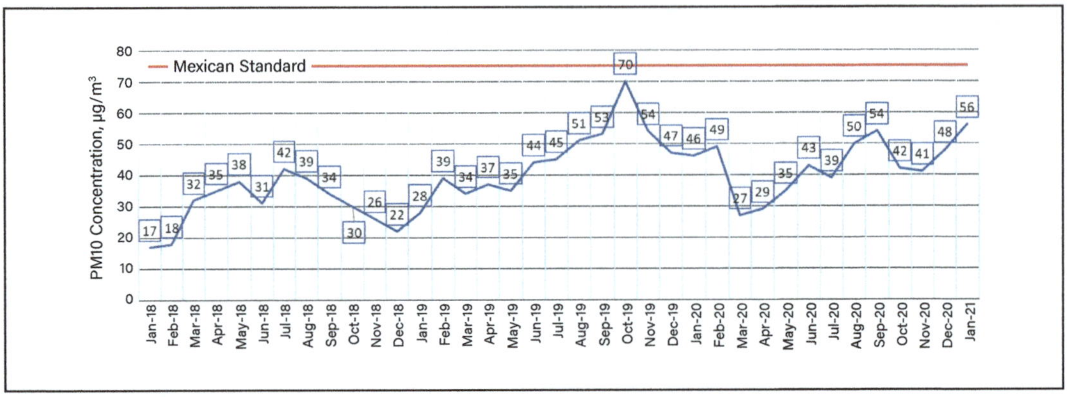

Courtesy of Newmont

FIGURE 3 Monthly average PM_{10} concentration from January 2018 to January 2021

will have a smoother passage. In addition, improving the quality of life for the community members living near mining operations helps build a strong relationship between community members and the mining company and is important as indicated in the Global Industry Standard on Tailings Management (GTR 2020).

In summary, the key to successfully managing dust at a TSF is ensuring that roles and responsibilities are clearly defined, the planning and sequencing process includes dust control, and budgeting for the associated costs related to dust control and management. Implementing these measures at the Peñasquito TSF resulted in sustainable improvements of dust control. ■

REFERENCE

GTR (Global Tailings Review). 2020. *Global Industry Standard on Tailings Management.* https://globaltailingsreview.org/wp-content/uploads/2020/08/global-industry-standard_EN.pdf. Accessed December 2020.

PART V

Lessons Learned from Failures

Stava Tailings Basins Failure, Stava, Italy

James L. Withiam

In the mountainous and steep valleys of northern Italy, the Stava tailings dam had retained waste materials from active fluorite mining operations since the early 1960s (Figure 1). Then on July 19, 1985, the impoundment, comprised of stacked upper (Figure 2A) and lower (Figure 2B) basins with embankments, catastrophically failed. Both embankments were then nearly 25 m high with outer slopes ranging between 1.5H and 1.2H (horizontal) to 1V (vertical) (34–40 degrees). The failure released approximately 240,000 m^3 of liquefied tailings that traveled downhill at velocities approaching 90 km/h, destroying the village of Stava and damaging Tesero, causing the deaths of 268 people, and covering an area of 435,000 m^2 up to 40 cm deep within a distance of 4.2 km below the impoundment.

The impoundment was constructed from 1961 to 1978 and was then idled until mining resumed in 1982. Both embankments were begun by starter dams, using available materials excavated from behind the starter dams. The embankments were raised by upstream construction using a pair of movable cyclones, which separated and deposited the heavier sand materials to form the outer shell, while the finer-grained materials were deposited into the basin as a slurry from various points along the ever-rising embankment crests.

In 1982, new ownership took over the dam, implementing significant changes compared to the previous operation practices

Courtesy of Fondazione Stava, 1985

FIGURE 1 Caretaker house and lower tailings basin, 1981–1982

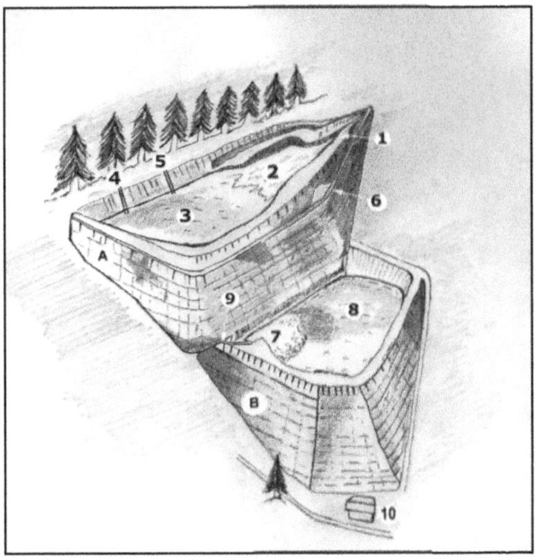

Source: Luino and DeGraff 2012

FIGURE 2 Tailings basin prior to failure (see text for explanation; 10 = caretaker house)

Source: Courtesy of Provincia Autonoma di Trento

FIGURE 3 Upper basin failure, 1985

(Figure 2). The changes included using a single, fixed cyclone (1) to deposit sand (2) and finer-grained material (3), trucking and dumping sand and gravel along the crest and service road (6), and, later, abandoning the cyclone and discharging from a single point at the rear of the basin. The service and emergency pipelines (4 and 5) that conveyed tailings and water from the upper to the lower basin became submerged in lower-basin deposits (7 and 8), and leakage (9) was reported from the base on the upper basin. Changes in operations gradually changed the deposition of sand (7) and finer-grained materials (8), location of the saturation zone, and stability of the outer shell and the impoundment.

Investigations of the failure concluded that many operating practices likely led to the collapse of the embankments (Figure 3) and the resulting deaths and destruction, although four things stood out (Decarli et al. 1985):

1. Excessive height and steepness of the impoundments
2. Upstream method of construction
3. Incorrectly installed drainage pipes
4. Change in disposal operations beginning in 1982, which altered the structural and hydraulic characteristics of the materials in the outer shell, driving the phreatic surface closer to the face of the impoundment ■

REFERENCES

Decarli, D., Giovannini, A., and Degasperi, A. 1985. *Tesero 19 Luglio 1985, Per Non Dimenticare* [Tesero, July 19, 1985, Never Forget]. Italy: Casa Editrice Publilux [Publilux Publishing House].

Luino, F., and DeGraff, J. 2012. The Stava mudflow of 18 July 1985 (northern Italy): A disaster that effective regulation might have prevented. *Natural Hazards and Earth System Sciences* 12:1029–1044.

Omai Tailings Embankment Failure, Guyana

Michael James
Agnico Eagle Mines

Construction of Tailings Dam No. 1 at the Omai gold mine in Guyana started in the early 1990s and continued until its failure. It was a rock-fill dam with a sloping clay core. A thin layer of filter material (sand) was placed between the clay core and the rock fill to prevent erosion of the core. At the time of its failure, the dam had a maximum height of 44 m. A corrugated steel culvert had been incorporated into the starter dike for the diversion of Captain Mann Creek during subsequent construction. After the completion of the starter dike, an upstream portion of the diversion conduit was plugged with cement grout. The impoundment created by the embankment was used for the storage of cyanide-laden tailings and wastewater (Haile 1997). Several aspects of the design and construction of the tailings storage facility embankment are notable:

- The criteria for the filter with respect to the adjacent rock fill was not adequately specified in accordance with the state of the practice. Specifically, the filter was undersized and not compatible with the grain size distribution of the rock fill.
- The filter did not always meet the criteria specified and was placed in a manner that promoted segregation.
- The diversion culvert was damaged by heavy equipment during construction.
- Seepage collars were not installed around the culvert.
- To avoid damaging the culvert, the soil (clay for the first 90 m from the upstream toe and then filter material) surrounding it was not compacted with the same effort as the soil in other zones and in some areas was not compacted.
- Five 1-m-wide soil-bentonite seepage cutoffs were constructed around the culvert to block seepage but were poorly compacted.
- The cement plug placed in the culvert was limited to an approximately 30-m-long upstream segment of the 150-m-long conduit. The remainder was left open.
- There was no redundancy in the design with respect to internal erosion.
- Inspection failed to rectify any of the obvious flaws in the design or correct inadequate construction practices.

On the night of August 19, 1995, excessive seepage was observed from the southern portion of the embankment. Subsequently, longitudinal cracks with vertical and horizontal offsets were observed on the crest and excessive seepage was observed from the northern portion of the embankment. Within 12 hours of the first observation of excessive seepage, the water level in the impoundment dropped 12 m. In total, approximately 3 to 4 million m^3 of cyanide-laden tailings and wastewater were released into downstream areas, mainly the Omai River (Haile 1997) (Figure 1).

Subsequent investigation revealed that erosion of the sand filter into the rock fill led to the creation of substantial voids and the collapse of portions of the clay core. Sinkholes and other features indicative of subsidence developed as far as 150 m from the conduit (Haile 1997).

The reviewers for the Guyana Geology and Mines Commission determined that the design and construction of the embankment, particularly the inadequacy of the sand filter and the poor compaction of the soil and of the soil-bentonite cutoffs around the culvert, allowed massive internal erosion leading to collapse of the clay core and failure of the embankment (Vick 1995). Additionally, the possibility that the culvert might have been damaged and provided a favorable pathway for the transport of water and soil through the embankment was deemed likely.

FIGURE 1 A team observing the cracked core at the Omai tailings embankment shortly after failure, 1995

Modifications to the design and construction could have averted the failure, based on general recommendations in USBR (1977) and the state of the practice:

- The filter should have been designed based on widely known filter criteria.
- The filter should have been placed properly to avoid segregation.
- The culvert should have been provided with seepage collars; the material around it should have been adequately compacted; and the entire length of the culvert should have been filled with cement.
- Given the great disparity between the grain size of the clay core and rock fill, it is likely that an adequate filter design would have required the use of a double filter, or at least a transition zone between the filter and the rock fill. ■

REFERENCES

Haile, J.P. 1997. Discussion on the failure of the Omai tailings dam. *Geotechnical News* 15:44–49.

USBR (U.S. Bureau of Reclamation). 1977. *Design of Small Dams*. Washington DC: USBR.

Vick, S.G. 1995. *Preliminary Report on Technical Causation—Omai Tailings Dam Failure*. Georgetown, Guyana: Dam Review Team, Guyana Geology and Mines Commission.

Geological Constraints on the Cadia Slump and Implications for Site Characterization, Cadia Valley Operations, Australia

Ian Gordon
Newcrest Mining Limited

John Wates
John Wates Consulting

Following the March 9, 2018, slump of the Northern Tailings Storage Facility (NTSF) southern embankment at Cadia Valley Operations (CVO), geotechnical investigations were commissioned by an independent technical review board (ITRB) charged with identifying the technical cause(s) of the slump. The investigations identified a low-density paleo-soil underlying Tertiary basalt immediately to the west of the slump, and it is proposed that these soils would have daylighted in the original topography beneath the slump. X-ray diffraction (XRD) analysis was used to explain the origin of the low-density paleo-soils, while paleogeographic reconstructions were prepared to explain their evolution and distribution. The low-density paleo-soil, which was relatively weak, highly compressible, and exhibited a high degree of brittleness, was found to constitute the most significant feature that contributed to the slump at this location. Key learnings from the Cadia ITRB investigations are discussed in the context of site characterization of tailings storage facilities (TSFs), a requirement of the Global Industry Standard on Tailings Management (GISTM; GTR 2020).

INTRODUCTION

CVO, a gold and copper mine owned and operated by Newcrest Mining Limited, is located 25 km south of the city of Orange, central west New South Wales, Australia. The mine is the largest block cave operation in Australia.

Late in the afternoon on March 9, 2018, a 300-m section of the southern embankment of the NTSF slumped, causing loss of containment of the stored tailings. The slumped embankment materials and tailings (~1.3 Mm^3 [million cubic meters]) were wholly contained in the Southern Tailings Storage Facility, located immediately downstream of the NTSF, and through the timely and perceptive evacuation of the area, there were no injuries or loss of life.

This case study focuses on the geological constraints on the slump, the geotechnical investigations undertaken to determine the cause of the slump, and key learnings for site characterization of TSFs.

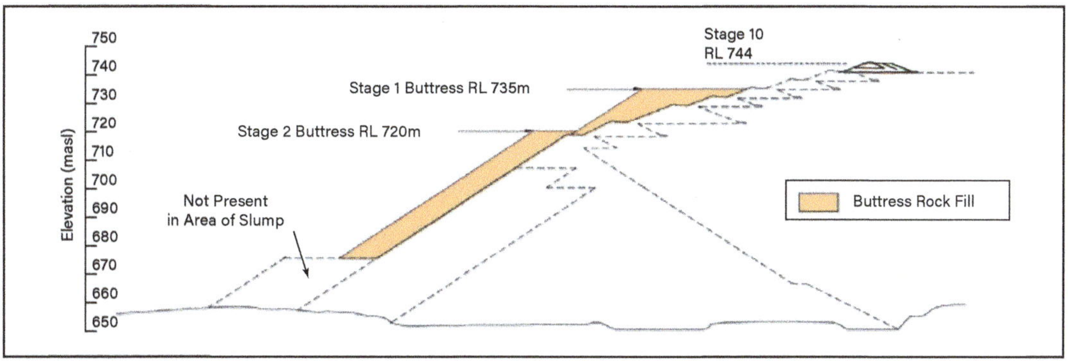

Source: Jeffries et al. 2019

FIGURE 1 Cross section of Northern Tailings Storage Facility southern embankment with proposed Stage 1 and 2 buttresses

BACKGROUND

Stage 1 of the NTSF, built using earth-cored rock-fill construction, was completed to a height of 50 m in 1998. The initial stage was followed by two raises using downstream construction, a single raise using centerline construction, and seven raises using upstream construction. At the time of the slump, the maximum height of the NTSF was 94 m.

A review of the tailings strength in 2017 indicated an overall strength profile lower than that used in the analysis of previous embankment stages and a long-term static factor of safety (FOS) that was too low. Consequentially, two buttresses were proposed to achieve acceptable FOS under both static and seismic loading (Figure 1). An upper buttress (Stage 1 buttress) was designed to increase the static short-term (undrained) FOS of a potential slip surface extending through the upstream raises, and a lower buttress (Stage 2) was designed to increase the static long-term (drained) FOS of a potential slip surface extending through the foundation. A sequence for construction of the buttresses was not prescribed, but a decision was made to proceed with Stage 1, prior to Stage 2, because this provided the largest FOS increase. Up to the time of NTSF embankment failure, the Stage 2 buttress had not been placed at the slump location.

THE SLUMP

Cracking at the crest level of the Stage 1 buttress was first noticed early on the morning of March 9. During the day, tension cracks developed over a length of 300 m parallel to the crest; the cracks became wider, and heaving was observed at the toe of the embankment. As the site had been evacuated late in the afternoon, the slumping event was not observed but occurred sometime between 4:30 and 7:00 PM and resulted in a displacement about 170 m to the south (Figure 2).

The ITRB (Jeffries et al. 2019), engaged to investigate the technical cause(s) of the slump, concluded through detailed investigation and analysis that the failure evolved in two phases.

Phase 1 involved all of the precursors of movement up to the time of evacuation of the site on March 9, 2018. Deformation analysis indicated that movement was initially small and in response to strain weakening of the foundation material and most likely commenced circa 2012 and continued through subsequent stages of embankment construction. Satellite monitoring data (InSAR, which stands for interferometric synthetic aperture radar) obtained after

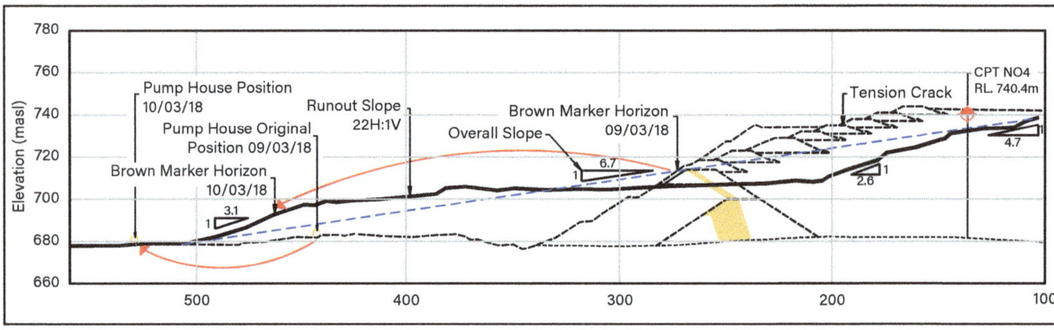

Source: Jeffries et al. 2019

FIGURE 2 Section through slump on March 10, 2018

the slump show an increased rate of movement in the months preceding the slump, which correlated with the Stage 1 buttress construction.

Phase 2 involved a rapid acceleration of movement that culminated in the slump features presented in Figure 2. Conclusions from the ITRB analysis were twofold:

1. Rapid acceleration of movement resulted from the substantial driving force created by static liquefaction of the saturated contractive tailings retained by the embankment.
2. Static liquefaction was triggered by the rapid yielding of the embankment foundations during the Stage 1 buttress construction and the reduction in lateral support offered to the tailings by the embankment.

GEOLOGICAL CONSTRAINTS ON THE SLUMP

The predominant geological unit near the NTSF is the Forest Reef Volcanic (FRV), comprising mainly andesitic volcanic rocks of Late Ordovician to Early Silurian age (~420 Mya). An irregular landform developed over the FRV was subsequently unconformably overlain by Tertiary basalt (~10 Mya), which has locally preserved, underlying zones of alluvium (paleo-alluvium), and soil profiles (paleo-soils) developed over the FRV. The Tertiary basalt has been largely removed by weathering and erosion with remnants remaining on ridge crests. One such remnant is present on the low ridge immediately to the west of the slump.

At the time of the initial investigations for the NTSF in 1995, the geological sequence at the embankment site, including the presence of paleo-alluvium, had been established. However, because of the small diameter of coring (~50 mm), undisturbed samples of soil strata beneath the basalt were not recovered. Similarly, where the FRV was not overlain by basalt, undisturbed samples were not taken from the residual soil profile intersected in drill holes or test pits, and the strength of foundation strata was based on triaxial tests completed on compacted disturbed samples.

The slump occurred where the embankment was ~56 m high, less than the maximum embankment height (94 m). It soon became apparent that this location was aligned with the eastern edge of the Tertiary basalt, and the underlying paleo-alluvium was initially considered to have contributed to the slump.

Under the guidance of the ITRB, an investigation program was initiated that included large-diameter sonic drilling, permitting high-quality undisturbed sampling below rock fill and the Tertiary basalt. Undisturbed block sampling was undertaken in residual soils and extremely weathered rock encountered in test pits. The field investigation program was supported by an extensive laboratory test program that included, among other tests, direct simple shear and triaxial tests on residual materials and critical state soil mechanics testing of tailings.

A major finding of the investigation was the presence of high-plasticity and low-density paleo-soil below the Tertiary basalt and paleo-alluvium (where present). Various theories were assessed to explain the presence of the low-density paleo-soil (faulting, FRV mineralogy, etc.); however, XRD analysis indicated that the clays within the paleo-soil were characteristic of weathering under a warmer, more humid paleoenvironment. Using this information, it was possible to distinguish between two FRV weathering products: Units A and B. Unit A has a significantly higher void ratio and lower density when compared with Unit B; but more importantly, Unit A is relatively weak, highly compressible, and strain weakening when subjected to load, exhibiting a high degree of brittleness.

A sequence of paleogeographic reconstructions of the NTSF area were prepared to explain the evolution and distribution of the low-density Unit A, with the final visualization presented as Figure 3. Where Unit A is overlain by Tertiary basalt, it is most likely that the basalt has prevented or inhibited the physical erosion of Unit A; but where the basalt is not present, it is most likely that Unit A has been eroded or modified to some degree under the current climatic regime exposing the underlying Unit B.

While the data are not abundant, it is proposed that Unit A would have daylighted immediately to the east of the Tertiary basalt (and beneath the slump) and would have been partially protected from physical erosion by the adjacent stronger basalt. Because of its proximity to

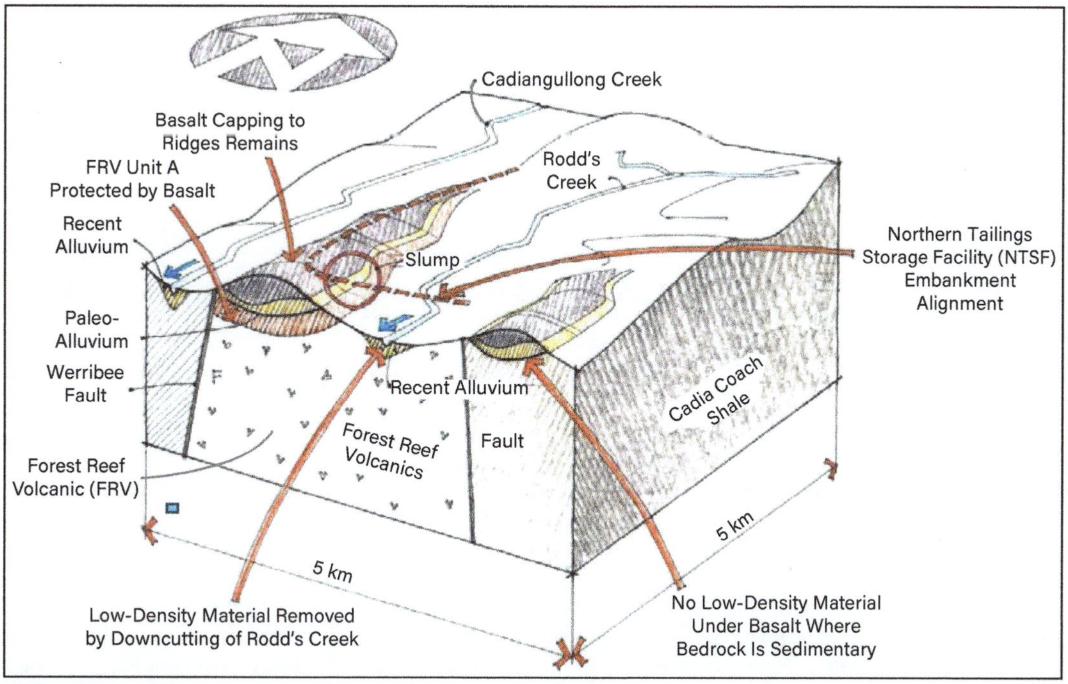

Source: Jeffries et al. 2019

FIGURE 3 Paleogeographic reconstruction showing FRV Unit A beneath slump and adjacent to the basalt

the foundation level, FRV Unit A was found to constitute the most significant feature that contributed to the slump at this location.

IMPLICATIONS FOR SITE CHARACTERIZATION

The GISTM (GTR 2020) details requirements regarding site characterization (Requirement 2.2) and the identification of brittle failure modes (Requirement 4.6). The identification of the geological constraints on the Cadia slump, undertaken as part of the ITRB investigations, is particularly relevant to the GISTM requirements and investigation methodologies outlined in guidelines such as the guidelines from the Association of Professional Engineers and Geoscientists of British Columbia (APEGBC 2016) and Australian National Committee on Large Dams (ANCOLD 2020).

Methodologies that proved key to defining the geological constraints on the Cadia slump, which may be equally applicable to the characterization of other sites include:

- The use of investigation techniques appropriate to the anticipated site conditions to obtain high-quality undisturbed samples for laboratory testing. This includes drill sizes and methods that allow appropriate undisturbed sampling and in situ testing of all materials, and seemingly time-consuming and labor-intensive methods, such as undisturbed block sampling of near-surface materials, which at the time of the investigations may appear too strong for more traditional sampling methods.
- Full characterization of all materials that have the potential to impact the design throughout the project life cycle.
- Focus on transition areas where materials may thin, change in weathering or strength, or move from one dominant property to another.
- Development of a conceptual geological model prior to the commencement of field investigations and continual assessment and refinement of this model throughout the evolution of the investigation and project development consistent with the methodologies initially outlined in the total geological/geomorphological approach of Fookes et al. (2001) and more recently developed as an International Association for Engineering Geology and the Environment publication (Baynes et al. 2022). ■

REFERENCES

ANCOLD (Australian National Committee on Large Dams). 2020. *Guidelines for Geotechnical Investigations of Dams, Their Foundations and Appurtenant Structures*. Hobart, Tasmania, Australia: ANCOLD.

APEGBC (Association of Professional Engineers and Geoscientists of British Columbia). 2016. *Site Characterization for Dam Foundations in BC:* APEGBC Professional Practice Guidelines, V1.2. Burnaby, BC, Canada: APEGBC.

Baynes, F.J., and Parry, S. 2022. *Guidelines for the Development and Application of Engineering Geological Models on Projects*. IAEG 25 Publication No. 1. Beijing, China: International Association for Engineering Geology and the Environment (IAEG). https://iaeg.info/C25EGMGuidelines/. Accessed November 2023.

Fookes, P.G., Baynes, F.J., and Hutchinson, J.N. 2001. Total geological history: A model approach to understanding site conditions. *Ground Engineering* (March): 42–47.

GTR (Global Tailings Review). 2020. *Global Industry Standard on Tailings Management*. https://globaltailingsreview.org/wp-content/uploads/2020/08/global-industry-standard_EN.pdf. Accessed November 2023.

Jeffries, M., Morgenstern, N.R., van Zyl, D., and Wates, J. 2019. *Report on NTSF Embankment Failure, Cadia Valley Operations*. Independent Technical Review Board Report prepared for Ashurst Australia. H356804-00000-22A-230-0001, Final. www.newcrest.com/sustainability/dams-and-tailings-management. Accessed November 2023.

Learnings and Outcomes from Córrego do Feijão Iron Ore Mine, Minas Gerais, Brumadinho, Brazil

David J. Williams
The University of Queensland

CASE STUDY/EXPERT REPORT FINDINGS

On January 25, 2019, the Córrego do Feijão iron ore mine tailings Dam I, located 9 km northeast of Brumadinho, in the State of Minas Gerais, Brazil, suddenly and catastrophically failed, killing 270 people, mainly workers. The expert panel investigating the technical causes of the flow liquefaction failure concluded that the sudden strength loss and resulting failure of the marginally stable embankment were caused by a critical combination of ongoing internal strains produced by creep, and a strength reduction generated by loss of suction in the unsaturated zone created by the intense rainfall toward the end of 2018 (Robertson et al. 2019). This followed several years of increasing rainfall after tailings deposition ceased in July 2016 (Figure 1). There was a disconnect between the wetting up of the transient low-permeability

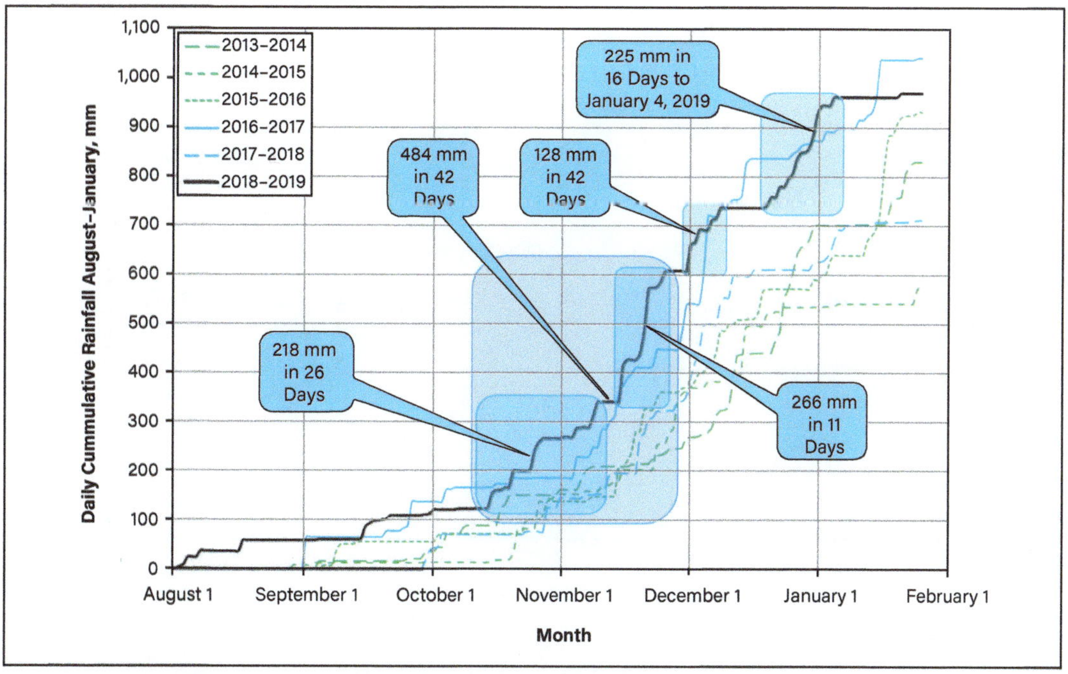

Source: Robertson et al. 2019

FIGURE 1 Increasing rainfall after tailings deposition ceased in July 2016

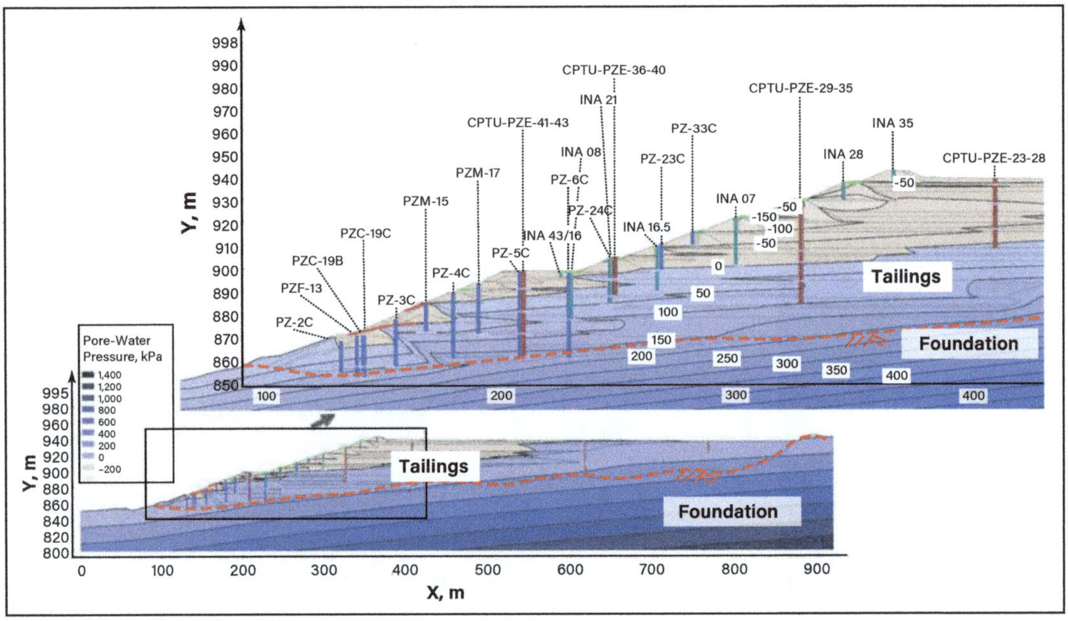

Source: Robertson et al. 2019

FIGURE 2 Extent of the unsaturated zone above phreatic surface

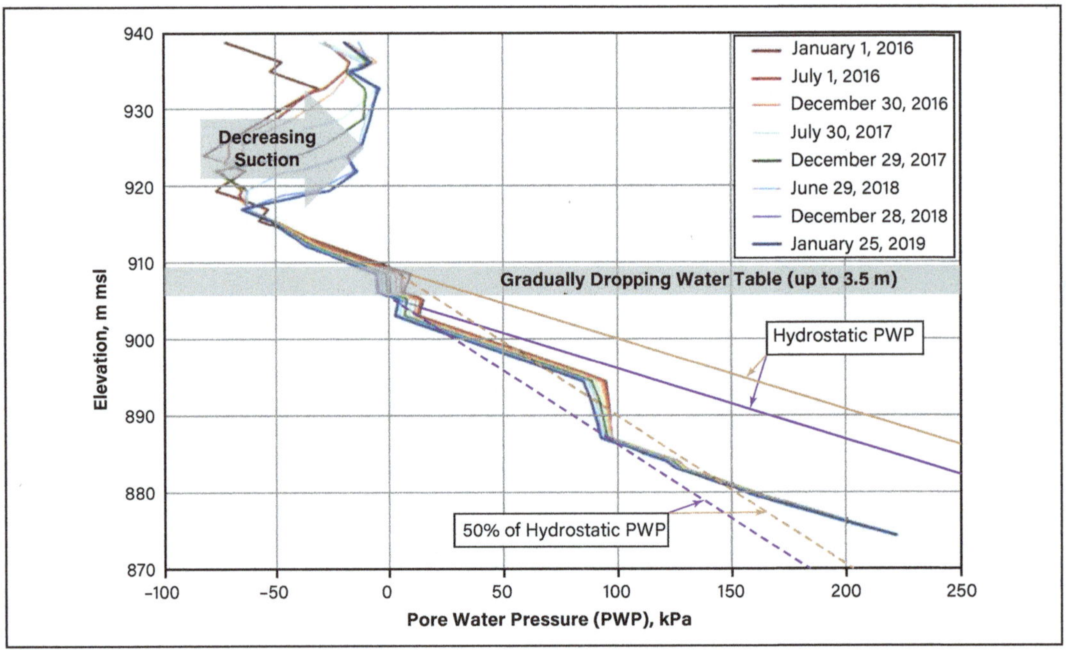

Source: Robertson et al. 2019

FIGURE 3 Loss of suction and hence shear strength of unsaturated zone within tailings deposit

unsaturated zone of the tailings deposit shown in Figure 2 and the essentially static saturated zone below the phreatic surface (Figure 3).

The Brumadinho tailings were bonded because of iron oxidation, making them brittle at low strain, and resulting in the tailings at different locations within the deposit being at different stress states. The failure appears to have initiated behind the setback in the downstream slope of the embankment, which located the upstream raises above the setback farther down the tailings beach, on wetter and softer tailings.

POSSIBLE CONSIDERATIONS IN RESPONSE

While there is no "silver bullet" to cover all tailings in all climatic, topographic, and seismic settings, much can be learned from recent high-profile and well-documented tailings facility failures, which, while rare, continue to occur at an unacceptably high frequency in terms of both industry and societal expectations (Williams 2020). A rethink is also required about the way in which many view tailings management in terms of economics, relying on a net present value accounting approach with a high discount factor, rather than a whole-of-life economics approach. There is scope for the further development and implementation of new tailings management technologies and innovations, and for the use of different economics models. Change is most readily achieved in new mining projects; hence, change in tailings management for the mining industry as a whole will be generational. ■

REFERENCES

Robertson, P.K., de Melo, L., Williams, D.J., et al. 2019. *Report of the Expert Panel on the Technical Causes of the Failure of Feijão Dam I*. www.b1technicalinvestigation.com/report.html. Accessed March 2021.

Williams, D.J. 2020. The role of technology and innovation in improving tailings management. In *Towards Zero Harm: A Compendium of Papers Prepared for the Global Tailings Review*. Edited by D. Brereton and A. Mihaylova. https://globaltailingsreview.org/wp-content/uploads/2020/09/GTR-TZH-compendium.pdf. Accessed January 2021. pp. 64–83.

Index

Note: *f.* indicates figure; *t.* indicates table.